W9-AFP-278

More Praise for

COUNT DOWN

"*Count Down* is that rarest of books: a compelling and engaging overview that not only rings the alarm bell but provides ideas for putting out the fire. Read this book if you want to protect your family's health."

—**Rick Smith, coauthor of** *Slow Death by Rubber Duck*

"Exposes the tacit bargain we've all struck. In exchange for the convenience of using in every aspect of our lives more and more plastics as well as non-FDA-approved chemicals—that's *most* chemicals, by the way—we've forfeited not just our own reproductive health but our children's. Swan lays bare this tragically bad deal in her powerful, page-turning *Count Down*. Read it."

—**Richard J. Jackson, MD, Director Emeritus,**
CDC National Center for Environmental Health

"Illuminates how our modern world is threatening our very existence. An eye-opening book that will leave you eager for individual and society-wide changes to begin *today*."

—**Nicole Avena, PhD, author of** *What to Eat When You Want to Get Pregnant*

"Compellingly readable ... a stirring call to action about the dangers posed by declining fertility, including the risks to our health, our economy, and even the future of the human race."

—**Jeremy Grantham, cofounder of the investment management firm GMO**
and the Grantham Foundation for Protection of the Environment

"Remarkable ... Swan illuminates the grave dangers posed by a class of manufactured chemicals called endocrine disruptors—which are produced each year in the millions of tons and incorporated into innumerable consumer products.... A powerful book whose message *must* be heeded by policy makers and the public—before it is too late."

—**Philip J. Landrigan, MD, MSc, founding director of**
Boston College's Global Public Health program

"Scrupulously illuminates the vast control that reproductive hormones have over matters sexual and the role that endocrine-disrupting chemicals play in undermining it.... This book should inspire all who read it to insist on EDC testing and regulations that quickly restructure chemical commerce into a form we can all live with."

—**Terrence J. Collins, Teresa Heinz Professor in Green Chemistry,**
Carnegie Mellon University

"Eloquent ... Reveals that humans are now effectively becoming an endangered species.... Dr. Swan offers important recommendations to counter our declining fertility that we'd all do well to follow."

—Bruce Blumberg, PhD, professor at the University of California, Irvine,
and author of *The Obesogen Effect*

COUNT DOWN

How Our Modern World Is
Threatening Sperm Counts,
Altering Male and Female
Reproductive Development,
and Imperiling the Future
of the Human Race

Shanna H. Swan, PhD

with Stacey Colino

Scribner

New York   London   Toronto   Sydney   New Delhi

Scribner
An Imprint of Simon & Schuster, Inc.
1230 Avenue of the Americas
New York, NY 10020

Note to the readers: The names and certain identifying characteristics of those who shared their personal stories have been changed.

First Scribner hardcover edition February 2021

SCRIBNER and design are registered trademarks of The Gale Group, Inc., used under license by Simon & Schuster, Inc., the publisher of this work.

For information about special discounts for bulk purchases, please contact Simon & Schuster Special Sales at 1-866-506-1949 or business@simonandschuster.com.

The Simon & Schuster Speakers Bureau can bring authors to your live event. For more information or to book an event, contact the Simon & Schuster Speakers Bureau at 1-866-248-3049 or visit our website at www.simonspeakers.com.

Interior design by Wendy Blum

Printed and bound by CPI Group (UK) Ltd, Croydon, CR0 4YY

3   5   7   9   10   8   6   4

Library of Congress Cataloging-in-Publication Data

Names: Swan, Shanna H., author. | Colino, Stacey, author.
Title: Count down : how our modern world is threatening sperm counts, altering male and female reproductive development, and imperiling the future of the human race / by Shanna H. Swan, PhD with Stacey Colino.
Identifiers: LCCN 2020040916 (print) | LCCN 2020040917 (ebook) | ISBN 9781982113667 (hardcover) | ISBN 9781982113674 (paperback) | ISBN 9781982113681 (ebook)
Subjects: LCSH: Male reproductive health. | Generative organs, Female. | Fertility. | Human reproduction. | Men—Sexual behavior. | Women—Sexual behavior.
Classification: LCC QP251 .S943 2021 (print) | LCC QP251 (ebook) | DDC 612.6—dc23
LC record available at https://lccn.loc.gov/2020040916
LC ebook record available at https://lccn.loc.gov/2020040917

ISBN 978-1-9821-1366-7
ISBN 978-1-9821-1368-1 (ebook)

For our children and grandchildren

Contents

Note to the reader: a glossary of acronyms and technical terms can be found at the back of this book.

PROLOGUE

It's hardly a news flash that human beings often take things for granted. Fertility is no exception—unless people discover they have a problem in this area. As with having access to basic necessities and certain fundamental freedoms, many people take it as a given that they'll be able to have babies when the time is right and help perpetuate the species. All of these assumptions reside under the notion that we don't always appreciate what we've got till it's gone, as folk singer/songwriter Joni Mitchell suggested in her hit song "Big Yellow Taxi."

It's difficult enough for a man or a woman, when experiencing reproductive disorders or fertility troubles, to accept that he or she may not be able to have children. Now there's an even greater challenge as human beings collectively are forced to contend with some dismaying biological realities. In Western countries, sperm counts and men's testosterone levels have declined dramatically over the last four decades, as my own research and that of others has found. Also, increasing numbers of girls are experiencing early puberty, and grown women are losing good-quality eggs at younger ages than expected; they're also suffering more miscarriages. It's no longer business as usual when it comes to human reproduction.

Other species are suffering, too. There's been a rise of abnormal genitals in wildlife, including unusually small penises in alligators, panthers, and mink, as well as an increase in fish, frogs, birds, and snapping turtles that have both male and female gonads or ambiguous genitalia. At first glance, these issues may seem like bizarre anomalies or cruel tricks from Mother Nature—but they're all signs that something very wrong is happening in our midst. Exactly what that culprit is continues to be hotly debated, but evidence pointing to likely suspects is mounting on a regular basis.

This much is clear: The problem isn't that something is inherently wrong with the human body as it has evolved over time; it's that chemicals in our environment and unhealthy lifestyle practices in our modern world are disrupting our hormonal balance, causing varying degrees of reproductive havoc that can foil fertility and lead to long-term health problems even after one has left the reproductive years. Similar effects are occurring among other species, adding up to widespread reproductive shock. Simply put, we're living in an age of reproductive reckoning that is having reverberating effects across the planet.

If these alarming trends continue unabated, it's difficult to predict what the world will look like in a hundred years. What does this dramatic decline in sperm count portend if it stays on its current trajectory? Does it signal the beginning of the end of the human race—or that we're on the brink of extinction? Does the environmental emasculation of wildlife suggest that the earth really is becoming much less habitable? Are we on the verge of experiencing a global existential crisis?

These are good questions, and we don't have clear answers to them, at least not yet. But pieces of the puzzle are being put together, as you'll see in the chapters that follow. You'll learn more about the breadth of these scary declines in sperm counts and other aspects of reproductive function, as well as the factors that are likely to blame for these unfortunate effects in human beings and other species, based on scientific research.

The following is clear: The current state of reproductive affairs can't continue much longer without threatening human survival. Current levels of sperm counts and concentrations, and decreased fertility, are already posing serious threats to Western populations, on both ends of the human life span: infertility is linked to an increased risk of certain diseases and earlier death in both men and women, while leading to a decrease in the number of children born over time. Obviously, this isn't a healthy scenario for *Homo sapiens* (or for other threatened or endangered species). Already, some countries with problematic age distributions are grappling with shrinking populations, with increasing numbers of older people being supported by fewer younger people.

It's a fairly bleak picture, I admit. But it's an important one to be aware of because, unless we take steps to reverse these harmful influences, the planet's species are in grave danger. Right now, the important measures that might improve the situation aren't happening. The 2017 publication of my meta-analysis on sperm-count decline in Western countries put this issue on the radar screen, grabbing headlines and television coverage around the world. But the findings haven't translated into committees being formed, environmental policies being changed, safer chemicals being manufactured, or other concerted efforts being made to address the suspected causes or protect our collective future.

Some people are in denial about the reality and gravity of the issue, and others shrug it off, saying the earth is overpopulated. Others acknowledge the sperm-count decline and the likelihood of a stagnation or decline in global population in the near future, but even they don't engage in much more than hand-wringing. In some ways, the sperm-count decline is akin to where global warming was forty years ago—reported upon but denied or ignored. Sometime between the 2006 release of Al Gore's Oscar-winning documentary *An Inconvenient Truth* and now, the climate crisis has been accepted—at least, by most people—as a real threat. My hope is that the same will happen with the reproductive turmoil that's upon us. Increasingly, scientists are in agreement on the threat; now, we need the public to take this issue seriously.

As a leading researcher on reproductive health and the environment, I feel it's my duty to draw attention to these alarming changes to sexual development and function. My interest in the effects of environmental factors on reproductive health started in the 1980s when I investigated a cluster of miscarriages in Santa Clara County, California, a trend that was eventually tied to toxic waste from a semiconductor plant that had leaked into the community's drinking water. Gradually, I became increasingly interested in investigating the potential effects that environmental chemicals can have on reproductive, sexual, and gender-related development, in men, women, and children. Over the last thirty years, I've conducted studies on everything from the origins of genital anomalies in newborns and the influence of prenatal stress on reproductive development in offspring, to the effects of many hours of TV watching on testicular function, the connection between high exposure to chemicals called phthalates and low interest in sexual activity, and many other subjects related to reproductive health.

Reversing the various reproduction-sabotaging effects that we're living with will require fundamental changes, including sweeping modifications to the kinds and volumes of chemicals that are manufactured and pumped into the environment. To make this happen, significant political and economic challenges will need to be overcome, a prospect that's daunting but urgently needed, in my opinion. Still, I believe this can be accomplished.

That's where this book comes in. In Part I, you'll learn more about the changes that are happening to reproductive and sexual development among humans and other species. Part II takes a detailed look at the sources of these shifts—namely, the environmental, lifestyle, and sociological factors that are contributing to these trends—and Part III explores the ripple effects the shifts are having on long-term health and survival. In Part IV, I will guide you toward smart ways to protect yourself and your unborn children as well as other steps you can take to help remedy what threatens both human and animal species. It's time to get started on altering these alarming trajectories and taking back the future. Consider this a clarion call for all of us to do what we can to safeguard our fertility, the fate of mankind, and the planet.

Part I

The Changing Landscape of Sex and Fertility

1

Reproductive Shock:
Hormonal Havoc in Our Midst

The Spermageddon Scare

In late July 2017, it seemed as if every media outlet around the globe had become obsessed with the state of human sperm counts. *Psychology Today* cried, "Going, Going, Gone? Human Sperm Counts Are Plunging," while the BBC declared, "Sperm Count Drop Could Make Humans Extinct," and the *Financial Times* announced, "'Urgent Wake-Up Call' for Male Health as Sperm Counts Plummet." A month later, *Newsweek* published a major cover story on the same subject: "Who's Killing America's Sperm?"

By the end of the year, my scientific paper "Temporal Trends in Sperm Count: A Systematic Review and Meta-Regression Analysis," which sparked these stories—and hundreds of others around the world— was ranked number 26 among all referenced scientific papers published worldwide, according to Altmetric's 2017 report.

This truly was the drop heard round the world.

These days, the world as we've known it feels as though it's changing at warp speed. The same could be said for the status of the human race. It's not only that sperm counts have plummeted by 50 percent in the last forty years; it's also that this alarming rate of decline could mean the human race will be unable to reproduce itself if the trend continues. As my study collaborator Hagai Levine, MD, asks, "What will happen in the future—will sperm count reach zero? Is there a chance that this decline would lead to extinction of the human species? Given the extinction of multiple species, often associated with man-made environmental disruption, this is certainly possible. Even if there is low probability for such a scenario, given the horrific implications, we have to do our best to prevent it."

This is especially worrisome because the sperm-count decline that's occurring in Western countries is unabating; it's steep, significant, and continuing, with no signs of tapering off. As Danish researcher and clinician Niels Skakkebaek, MD, who was the first person to alert the scientific community to the role of environmental factors in sperm decline, said, "It's an inconvenient message, but the species is under threat, and that should be a wake-up call to all of us. If this doesn't change in a generation, it is going to be an enormously different society for our grandchildren and their children." Indeed, if the decline continues at the same rate, by 2050 many couples will need to turn to technology— such as assisted reproduction, frozen embryos, even eggs and sperm that are created from other cells in the laboratory (yes, this is actually being done)—to reproduce.

A Dystopian Future?

Some of what we've been thinking of as fiction, from stories such as *The Handmaid's Tale* and *Children of Men*, is rapidly becoming reality. In the winter of 2017, I presented my sperm-decline findings at the One Health, One Planet conference, which focused on the interconnected health of different species on the planet, the damage being inflicted by our mad "industrialization" of the environment, and its devastating effects on frogs, birds, polar bears, and other species. After presenting the results of our analysis, which were shocking enough to the audience, I spoke for the first time about what sperm decline could mean for *Homo sapiens*. That night, I awoke from a dream, feeling incredibly anxious as

I suddenly realized the full implications of the story I'd put together—
that given the declines in sperm count and testosterone levels and the
increases in hormonally active chemicals that are being spewed into the
environment, we really *are* in a dangerous situation for mankind and
world fertility.

This was no longer only a matter of scientific study for me. I felt and
remain genuinely *scared* by these findings on a personal level.

In some ways, the picture looks even worse when you delve deeper
because it's not just an issue for men. Women, children, and other species
are also having their reproductive development and function comman-
deered in a dysfunctional direction. In some countries throughout the
world, including the United States, a massive sexual slump is underway, due
to declines in people's sex drives and interest in sexual activity; men, including younger guys, are also experiencing greater rates of erectile dysfunction.
In animals, there have been changes in mating behavior, with more reports
of male turtles humping other male turtles, and female fish and frogs
becoming masculinized after being exposed to certain chemicals.

Taken together, these trends are causing scientists and environmentalists to wonder, How and why could this be happening? The answer is
complicated. Though these interspecies anomalies may appear to be distinct and isolated incidents, the fact is that they all share several underlying causes. In particular, the ubiquity of insidiously harmful chemicals in
the modern world is threatening the reproductive development and functionality of both humans and other species. The worst offenders: chemicals
that interfere with our body's natural hormones. These endocrine-
disrupting chemicals (EDCs) are playing havoc with the building blocks
of sexual and reproductive development. They're everywhere in our
modern world—and they're inside our bodies, which is problematic on
many levels.

Here's why: Hormones—particularly, two of the sex hormones,
estrogen and testosterone—are what make reproductive function possible.
Both the amount of each hormone and the ratio between these hormones
are important for both sexes. The sweet spots for these ratios are different

for each sex: depending on whether you are a man or a woman, your body needs optimal amounts of estrogen and testosterone, not too much or too little of either one. To make it more complicated, the timing of their release can alter reproductive development and functionality, and the transport of hormones can be an issue as well—if they don't get to the right place at the right time, essential processes such as sperm production or ovulation won't be set into motion. Endocrine-disrupting chemicals, as well as lifestyle factors—including diet, physical activity, smoking, and alcohol or drug use—can alter these parameters, sending levels of these crucial hormones in the wrong direction.

High-Altitude Worries

Another, no less important or complicated, question, is, What do these reproductive changes mean for the fate of the human race and the future of the planet? It's not just a matter of survival—whether humans will continue to be able to reproduce or whether the human race will die out in a *Children of Men*–type scenario. These issues have subtler, more personal consequences as well. Take declining sperm counts: statistically, this phenomenon goes hand in hand with many other problems for males, including an increased risk for cardiovascular disease, diabetes, and premature mortality (you'll learn more about these downstream health hazards in chapter 8).

And again, this isn't just about men. Not only is women's fertility being affected, even if less obviously or dramatically, but sperm quality can be altered by changes that occur when male fetuses are in the mother's womb. At that time the fetus is affected by the mother's choices and habits, which means that women can serve as conduits for potentially harmful chemical exposures. Contrary to previous belief, the womb does *not* protect the fetus against chemical assault, and a developing fetus has few defenses against the infiltration of chemicals. Looked at another way, the most important events in a male's life, in terms of sexual and reproductive development, occur while he's still in utero. Babies and children are more vulnerable to these chemical assaults than adults, but those who are most vulnerable haven't been born.

The sperm decline signals changes that affect everybody.

As some population experts and scientists put it, "a demographic time bomb" is on the horizon—future generations won't be able to meet the financial and caretaking needs of an ever-increasing number of older adults and retired workers, given the declining fertility rate. And the changes in sexual development taking place all over the world appear to have been accompanied by an apparent rise in gender fluidity,* which is not a negative development, in my opinion. The point is, human sexuality and society are in flux, and this flux affects us all. It's as if the snow globe has been shaken, altering the reproductive landscape inside—only this is happening in real life.

What comes to mind when you see a reference to the "1 percent effect," a common phrase in the cultural lexicon? Most people think of socioeconomic status, namely a ranking in the top 1 percent of wealth in the United States. Not me. I think of the fact that the rate of adverse reproductive changes in males is increasing by about 1 percent per year. This includes the rates of declining sperm counts and testosterone levels, increasing rates of testicular cancer, and the projected worldwide increase in the prevalence of erectile dysfunction. On the female side of the equation, miscarriage rates are also increasing by about 1 percent per year. *A coincidence?* I think not.

Questioning the Issues

If you're skeptical about all this, that's fair enough. I used to be, too. Whether it's because I'm a trained scientist or a natural-born skeptic, I've

---

* 　　　　Many countries are experiencing increases in issues related to gender identity, gender fluidity, and gender dysphoria. *Gender dysphoria* refers to the feeling that one's emotional and psychological identity as male or female is out of sync with one's biological sex. (You'll read more about this in chapter 4.)

always been a firm believer in Albert Einstein's assertion that "blind belief in authority is the greatest enemy of truth." That axiom has underscored all of my research on environmental influences on human health—including the effects of endocrine-disrupting chemicals, water contamination, and drugs—as well as my interpretation of other people's research. So when the *British Medical Journal* published a study in 1992 that claimed worldwide sperm counts had fallen significantly in the previous fifty years—which was a major bombshell—I found the issue intriguing, but I had significant doubts about the validity of the results.

After reading and rereading what came to be known as the Carlsen paper—named after lead author Elisabeth Carlsen—I was among the skeptics who questioned the methodology and the selection of samples, and I thought of many potential biases that might have distorted the findings. Granted, I was hardly alone; numerous critiques and editorials ensued. But the findings of that study were so important from a public health perspective that I couldn't put them out of my mind, even though I was busy doing research about the risk of birth defects and miscarriage from solvents in drinking water. Doubtful as I was about the findings of that particular study, I knew that certain environmental chemicals *could* be decreasing sperm counts, so I wanted to investigate; it felt like a bit of a detective case.

In 1994, I was appointed to the National Academy of Sciences Committee on Hormonally Active Agents in the Environment, and soon after, I was asked to tell the committee whether the Carlsen paper's conclusions were justified. For six months, I combed the literature to find all the criticisms that had been raised about the paper, then I reviewed the sixty-one studies the Carlsen team had included in its analysis to try to address those criticisms. Particular questions I pursued included: Did the early studies include healthier, younger men than the later ones did? Did the later studies include more smokers or obese men, which would create a distorted picture of what was happening? Had the method of counting sperm changed over fifty years in a way that made more recent sperm counts lower?

To get to the bottom of this mystery, I found two colleagues, Laura Fenster and Eric Elkin, who were willing to help me. The results were utterly astounding: after six months of data crunching and considering potential biases and confounding factors, our overall conclusion agreed, almost exactly, with that of the Carlsen team. Because we'd accounted for geographic location in the various studies, we found that sperm counts really *were* declining in the United States and Europe. But what about the rest of the world?

After these findings were published in 1997, I felt that we needed to ask whether sperm counts were different in different locations, since that would point to environmental factors at play. I've spent the last twenty years basically trying to answer that question. After conducting many more studies on semen quality, sperm decline, and related factors, I feel that I have. Not only have I shifted from being dubious to being utterly convinced that a dramatic decline in sperm counts is occurring, I've also discovered that various lifestyle factors and environmental exposures may be acting in tandem or in a cumulative fashion to fuel the decline.

Fast-forward to the summer of 2017 when my latest paper on this subject, written with my colleague Hagai Levine and five other committed researchers, went viral.

The news my colleagues and I reported in our meta-analysis: Between 1973 and 2011, sperm concentration (the number of sperm per milliliter of semen) dropped more than 52 percent among random men in Western countries; meanwhile, the total sperm count fell by more than 59 percent. We came to these conclusions after examining the findings from 185 studies involving 42,935 men that had been conducted during this thirty-eight-year period. To be clear: these men weren't selected based on their fertility status; they were everyday Joes and Johns, ordinary men.

Given that the findings pertain primarily to Western countries, this may sound like a first-world problem, but it's not. Rather, I suspect that societies in which people are likely to begin having children at a younger age are less likely to be affected by the fertility-damaging effects of environmental chemicals and life stressors. In our meta-analysis, there were much less data on sperm counts from men from South America, Asia, and Africa; however, more recent research reports declines in those regions as well.

Taking This Personally

What does all this mean in relatable terms? When people hear about these threats to their fertility, it's a big blow to their egos, their sense of potency, and their confidence in being able to sustain themselves as a family, a culture, and a species. It's startling and chilling when you realize that the number of children you may be capable of having is slightly less than half of that your grandparents could conceive. It's also shocking that in some parts of the world, the average twenty-something woman today is less fertile than her grandmother was at thirty-five.

The precipitous drop in sperm counts is an example of a "canary in the coal mine" scenario. In other words, the sperm-count decline may be Mother Nature's way of acting as a whistleblower, drawing attention to the insidious damage human beings have wrought on the built and natural worlds.

Which leads to a third, crucial question about all this: What can we do about the problem? There are steps we can take both as individuals and as a society to stay healthy and protect our sexual development. But the first thing we have to do is learn more about the nature of these problems. Most people outside the scientific community are totally unaware of these disturbing trends, and as a researcher who is committed to identifying environmental causes of reproductive health problems, I feel it's my duty to draw attention to them.

Whether it's through our lifestyles or the chemical contaminants we've brought into the world, we, as human beings, have inadvertently unleashed these problems. At this rate, it's hard to know what the future will look like, unless we take conscious and considered steps to protect ourselves and curb the chemicals that are infiltrating our daily lives. The time has come for us to stop playing Russian roulette with our reproductive capacities.

2

THE DIMINISHED MALE:

Where Have All the Good Sperm Gone?

A Date with Donation and Destiny

Mondays tend to be slow and quiet days at the Fairfax Cryobank in Philadelphia, especially compared to Fridays. On Fridays, men between the ages of eighteen and thirty-nine are often booked back-to-back for one of the two private rooms (where the recommendation is "Bring what you may need," as in porn) to engage in the act of sperm donation. There's a simple reason Mondays aren't as busy: men who are donating sperm are advised to abstain from sexual activity for seventy-two hours to set them up for an optimal sperm sample—abstinence affects the concentration and volume of a sperm sample—and not many men are willing to do that over the weekend. "We want to see good-quality specimens, and with about seventy-two hours of abstinence, most guys will have the best percentage of motile sperm," explains Michelle Ottey, PhD, laboratory director and director of operations at the Fairfax Cryobank. "Sometimes they have it and sometimes they don't. They're not always good at listening to our advice about the abstinence hours."

Sperm have always been a precious commodity, given the critical role they play in generating new life. Even a relatively small change in the typical sperm count has a substantial impact on the percentage of men who will be classified as infertile or subfertile. It's not just about the number of sperm, however; certain qualities, including the movement patterns, of these little swimmers are also essential for them to be able to wiggle upstream to meet the egg of their dreams.

After a man starts producing sperm during early adolescence, he's at continuous risk for potential harm to his swimmers, a vulnerability that lasts for the rest of his life. That's because spermatogenesis, the production of sperm, which occurs in the seminiferous tubules that form the bulk of each testicle, starts in early puberty (when a boy is ten to twelve years old) and continues throughout his life. In a healthy, fertile man, the testicles produce 200 million to 300 million sperm cells per day, only about 50 percent of which become viable sperm. It takes about sixty-five to seventy-five days for sperm to mature, and a new cycle of sperm production starts approximately every sixteen days. When the sperm mature, they leave the tubules and enter the epididymis, a coiled, tubular organ that's attached to the testicles.

Here, the mature sperm learn to "swim" and fine-tune their movement. Mature sperm resemble microscopic tadpoles: they have an enzyme-coated head, a tail, and a thinner portion of the tail, called an end piece. Once inside the epididymis, mature sperm wait to be ejaculated into the vagina (or somewhere else)—not unlike the scene depicted in Woody Allen's film *Everything You Always Wanted to Know About Sex*, where the sperm are waiting to "parachute" out of an aircraft and complete their mission. On average, each time a man ejaculates he releases two to six milliliters—about a teaspoon—of semen, which contains as many as 100 million sperm. Even the healthiest, best-shaped sperm don't pause to ask for directions; a relatively small percentage of sperm will swim in the right direction—as in, toward a beckoning egg. If the man doesn't ejaculate, the sperm will die and get reabsorbed by the body. The reality is, sperm tend to live fast and die young.

Sperm 101

The study of sperm began in a fairly bizarre fashion. In 1677, Antoni van Leeuwenhoek, a Dutch tradesman and self-taught scientist who was fascinated with microscopes, collected his semen after having sex with his wife and examined it under a microscope: He saw millions of tiny, wriggling shapes that he called animalcules (little animals) swimming in the fluid. He believed that each sperm contained a miniature, pre-formed human being that would unfurl and develop inside the mother after being nourished by the female egg.

That theory was obviously debunked long ago. But what van Leeuwen-hoek saw under the microscope is the same as what we see today when examining a magnified semen sample from a fertile man: A healthy sperm cell is made up of a torpedolike head that contains DNA, a middle section that's packed with energy-providing mitochondria, and a relatively long tail that propels the sperm forward. Each sperm is minuscule—roughly .05 millimeter or .002 inch long—much too tiny to be seen by the naked eye.

In the scientific world, research protocols often change over time, but when it comes to counting sperm, the method endorsed by the World Health Organization hasn't changed much since the 1930s. Sperm are still counted using the hemocytometer, an instrument that was invented in 1902 by French anatomist Louis-Charles Malassez and originally used to count blood cells. The device consists of a thick glass slide with a rectangular indentation that creates a chamber that contains a laser-etched grid of perpendicular lines. To evaluate a man's sperm concentration at a sperm bank or another lab, a drop of semen is placed on a slide and examined under a microscope, and a trained technician counts how many sperm are within a square on a grid pattern.

In human beings, normal sperm concentration* ranges from 15 million to greater than 200 million sperm per milliliter (or per mL) of semen. The World Health Organization has officially deemed a concentration of fewer than 15 million per mL to be "low." But according to a much-cited Danish study, men with a sperm concentration of less than 40 million per

---

*   Sperm count is an overarching term that refers to both sperm concentration and total sperm count. Sperm concentration is expressed as millions of sperm per milliliter, whereas total sperm count is equal to the sperm concentration times the volume of the ejaculate sample and is expressed as millions of sperm.

milliliter are considered to have an impaired likelihood of conceiving. (My own research found that in 1973 the average man in Western countries had a sperm concentration of 99 million per milliliter; by 2011, it had fallen to 47.1 million per milliliter. But we'll come back to that shortly.)

For fertility, it isn't just the number of sperm that matters; it's also about the sperm's shape and how they move. That is, are they able to swim in a way that suggests they're likely to be able to reach and penetrate an unfertilized egg? If sperm are swimming in a circle (what's called nonprogressive motility), that's not good; it's the equivalent of revving your car's engine in neutral—you're not going to get anywhere. If they're not moving at all, but instead are hanging out like couch potatoes with hangovers, that's a problem, too, since such immobility tends to persist. Sperm that move too slowly or sluggishly—with a forward progression of less than twenty-five micrometers per second—simply aren't going to get to their intended target.

What's considered normal or acceptable motility varies considerably among species. A man must have total sperm motility of greater than 50 percent to be considered normal in this respect; by contrast, to pass a soundness exam for breeding, stallions are recommended to have greater than 60 percent and dogs should have greater than 70 percent progressively motile sperm.

The parameters that are used to evaluate sperm quality under a microscope include *concentration* (how densely sperm are packed in a unit volume of semen); *vitality* (the percentage of sperm that are alive); *motility* (the sperm's movement or swimming ability); and *morphology* (the size and shape of sperm). All of these metrics matter, and based on recent evaluations of these elements, the *quality* of human sperm is going down as well as the quantity.

Aside from a complete absence of sperm (called azoospermia),* no

---

* Azoospermia can happen if the testicle is not producing any sperm or enough sperm to be detected in a standard semen analysis, or if sperm is produced but can't be discharged because of an obstruction.

single sperm parameter can predict that a man will be completely infertile, though they're all related to the chances of successfully achieving a pregnancy. The standard "big three"—sperm concentration, motility, and morphology—are routinely used to assess semen quality and fertility. Studies have found that when reproductive-medicine clinicians examined the three major measures of semen quality in approximately fifteen hundred men, a little more than half of whom were fertile and slightly less than half of whom were infertile based on these sperm parameters, all three parameters mattered in identifying the infertile men. But there was an additive effect: when any *one* of the measures was in the infertile range, the man was about twice as likely to be infertile as a man with none of these measures in the infertile range; when any *two* of the measures were in the infertile range, he was six times more likely to be infertile; and when all *three* measures fell in the infertile range, his chance of being infertile was sixteen times higher.

Giving at the Office

When a man donates to a sperm bank, his sperm need to meet certain benchmarks, only one of which relates to the sperm count. Sperm banks, whose specialty is, of course, collecting viable sperm in mass quantities, are facing mounting challenges across these different criteria. In a study published in 2016 involving 9,425 semen specimens from nearly five hundred men, researchers found a significant decline between 2003 and 2013 in sperm concentration, motility, and total count among young adult men who were attending or had recently completed college in the Boston area. While 69 percent of the aspiring sperm donors made the cut in 2003, only 44 percent did in 2013. This was true despite that the more recent group of guys had improved lifestyle variables such as a decline in alcohol use, smoking, and body weight and an increase in steady exercise.

Similarly, in a recent study involving potential sperm donors ages nineteen to thirty-eight throughout the United States, researchers examined more than one hundred thousand semen specimens and found a decline in total sperm count, sperm concentration, and motile sperm between 2007 and 2017. Downward trends are occurring in other countries, too. In China, for example, among young men who applied to be sperm donors at the Hunan Province Human Sperm Bank of China, the percentage of qualified donors dropped from 56 percent in 2001 to 18 percent in 2015, a two-thirds decline.

By any criteria, sperm just aren't doing well these days. And most men don't even realize this.

While the Fairfax Cryobank has experienced an increase in sperm donors in recent years, thanks to its expanded recruitment efforts, the sperm bank has seen a drop in sperm count and motility among freshly donated sperm samples. Before being suitable for use in intrauterine insemination (IUI) or in vitro fertilization (IVF), donated sperm must undergo a washing process, often involving centrifugal force—not to make the sperm shiny and polished for their big date with an egg but to remove chemicals, mucus, and nonswimming sperm from semen and to separate sperm from the seminal fluid. After the wash, sperm are placed in vials. "Since I started working here in 2006, we have seen a decrease in the number of vials per sperm sample that we're able to get—by about half," Dr. Ottey says. This is especially significant because most sperm samples are frozen for later use—"They literally are frozen in time"—and approximately 50 percent of the healthy, motile sperm cells that are collected in a sample and frozen won't survive the freeze-thaw process; they'll die.

Yet, while the supply of high-quality sperm is declining in some parts of the world, the demand for healthy, viable sperm has increased. The rising rates of abnormal and inadequate sperm volume are certainly playing a role, but another big driver is the uptick in requests from different demographic groups: in particular, more single women and same-sex couples are looking to have children—and they need high-quality sperm to achieve their goal. Prospective parents could use sperm from a friend or family member (often referred to as known donors)—and some do—but, for obvious reasons, this can be emotionally fraught. The other option is to use a strictly screened stranger's (an anonymous donor's) sperm through a sperm bank or fertility clinic—and that's where the demand is highest. In 2018, the global sperm-bank market was valued at $4.33 billion; it's expected to reach $5.45 billion by 2025. A widely touted estimate is that thirty thousand to sixty thousand children are conceived through sperm donation each year in the United States alone.

Playing the Infertility Blame Game

Why do these sperm supply-and-demand details matter? Because, beyond the doomsday scenarios that garner headlines, all too often the psychological and medical burdens of dealing with fertility issues have been placed squarely on women's shoulders. Not only is this incorrect on the most basic level—given that it takes viable sperm as well as a healthy egg to create a pregnancy—it's especially wrong now, when a high proportion of infertility issues can more clearly be placed at men's feet.

Admittedly, only recently have scientists and medical professionals begun to appreciate the extent to which fertility depends on the health and environment of *both* the male and the female partner, as well as the interactions between them. Historically, fertility has been a concept applied only to women. One reason is that demographers have traditionally defined the fertility rate as the average number of live births per woman of reproductive age. It's widely known that a woman loses precious eggs as she gets older, and as a result, constant reminders appear in the media and elsewhere about the worrisome ticking of women's biological clocks and the impact that certain lifestyle practices can have on fertility. Many women are aware of these realities, and some feel pressure to settle down and have babies by a certain age. *Men?* Not so much.

The recent decades have seen a substantial change in perspective, at least in the scientific community, as it has become increasingly recognized that men contribute to a greater proportion of infertility cases than previously believed. Male reproductive issues are currently thought to cause approximately one-quarter to one-third of infertility cases, equal to the proportion of female reproductive challenges. The remaining cases of infertility stem from a combination of male and female factors—perhaps a woman is slightly subfertile (because she has irregular ovulatory patterns, for example) and her male partner is also a bit subfertile (due to reduced sperm motility), so they have trouble conceiving. But if either was with a partner who was incredibly fertile (yes, some people really are), getting pregnant wouldn't be as challenging.

The Fertility Literacy Gap

Despite these realities, most men are unaware that the quality of their sperm can affect their chances of successfully conceiving a pregnancy. If they ejaculate plenty of semen, they think they're good to go, which isn't necessarily true. A 2016 Canadian study found that, while most of the 701 participating men considered themselves to be at least somewhat knowledgeable about male reproduction and fertility, many were unable to identify risk factors—such as obesity, diabetes, alcohol consumption, and high cholesterol—that are associated with male infertility.

In general, men have a no-problem attitude toward conception: they simply assume that if they want to have children, they'll be able to impregnate their partner easily enough. But that isn't always the case, especially in our modern world.

As an example, consider Megan and James, former multisport college athletes who are still physically fit: They believed it would be a cinch for them to get pregnant when they were ready to start a family. It wasn't. Megan, thirty-four, a nutrition consultant, and James, thirty-two, a banker, tried to conceive for a year without success, at which point they both began to question *her* fertility status. So Megan went to her ob-gyn and had a battery of physical examinations and blood tests that indicated everything seemed to be A-OK. When James subsequently went to a urologist for a comprehensive checkup, he discovered that his sperm count and motility rate were slightly low and that he had a narrowing of the pathway through which semen travels before being ejaculated. James felt blindsided by the news, especially because he'd always thought of himself as a superhealthy, virile man.

When the urologist asked about James's lifestyle habits, most of which were pristine, he learned that James would relax in a hot tub or steam room after playing squash or working out, four or five times per week. The urologist advised him to eschew these hot environments because severe heat is known to be toxic to sperm. After steering clear of these hot spots for several weeks, James and his wife conceived on their own. Naturally, they were thrilled, but James was left feeling flummoxed: How could he have had this sperm-flow problem all these years without knowing about it? Why hadn't anyone ever told him that frequent exposure to heat could harm his swimmers? "Women receive lots of information about how to prepare their bodies for making a baby—why don't men?" James asked.

As James discovered, it's not unusual for men to have no clue that there's a problem with their sperm or its transport system until they try to make a baby. This happened to Daniel, forty, and his wife, Laura, thirty-five, who spent a year trying to conceive, to no avail. After they both had tests done, Daniel was diagnosed as infertile because his sperm were abnormally shaped—few had all the component parts. This was at least partially caused by a condition called varicocele, an enlargement of the veins in the scrotum, which can decrease sperm count and reduce sperm quality.[*] "When the doctor said I would probably never have kids of my own, I was devastated," recalls Daniel, an attorney. "I still have no idea why or how I could have had this condition without knowing it." But he wasn't willing to give up hope, so he underwent a procedure to correct the varicocele, which improved his semen and sperm quality over the subsequent six months. The couple now have four-year-old twins.

---

[*]   BTW: A study involving more than 1.3 million teenage boys in Israel found that the incidence of varicocele more than doubled between 1967 and 2010 for reasons that have yet to be determined.

Down for the Count

Given the declines in sperm counts and other measures of sperm quality in Western countries, men's share of infertility cases may be on the rise. A recent study involving patients presenting for care at infertility centers in New Jersey and Spain found that the proportion of men with total motile sperm counts greater than 15 million per mL had declined approximately 10 percent between 2002 and 2017, which suggests a notable drop in sperm counts even among "subfertile men." An unfortunate form of double jeopardy, the implication is that subfertile men may be becoming even more subfertile.

The ratio of intracytoplasmic sperm injection (ICSI) procedures—which involve injecting live sperm directly into a human egg—to all IVF procedures has been increasing in many countries; this could suggest that male factor infertility is increasing, according to Danish researcher and clinician Niels Skakkebaek. The use of ICSI, available since 1991, has more than doubled from 1996 to 2012, among fresh IVF cycles in the United States. One of the major gifts ICSI has provided is that it has brought male factor infertility out of hiding and allowed it to be treated as a medical problem, rather than "a manhood issue."

Meanwhile, the World Health Organization's reference value for the lowest sperm concentration that's compatible with fertility—meaning, it takes less than a year for a man and his partner to achieve a pregnancy—has declined over the last thirty years. Clinicians tend to use this number as a cutoff when deciding whether to send a man for a complete fertility workup. The point is, our idea of what's a "good enough" sperm concentration has actually gone down. It used to be 40 million/mL, then it was lowered by the WHO to 20 million/mL in 1980 and to 15 million/mL in 2010. For the sake of comparison, back in the 1940s, 60 million/mL was considered an adequate sperm count.

These changes can have unintended consequences. On the upside, this lower cutoff eases the burden on fertility clinics and might make men with relatively low sperm concentrations—by previous standards—feel better.

But it doesn't do them any favors in terms of their fecundity. And if men are told their sperm concentrations are fine, they're more likely to wait until they're older to try to impregnate their female partners, and their older age could make it even harder for them to achieve a pregnancy.* While it's not widely acknowledged, women aren't the only ones to undergo an age-related decline in fertility. Several sperm parameters decline with advancing age, with the most marked changes being a loss of volume of sperm, a decrease in motility, and an increase in DNA fragmentation, the presence of abnormal genetic material within the sperm. Basically, declines in sperm quality *and* quantity make every aspect of fertility harder as men get older.

In recent years, the WHO has made similar reductions in the reference values for sperm motility, volume, vitality, and morphology. All of these factors are correlated with fertility: if a man has a low sperm count, he's more likely to have sperm that don't swim well or have the right shape. And keep in mind that, even in a best-case scenario, with a healthy adult man who has tens of millions of sperm per ejaculate, very few—perhaps only one in a million—will succeed in connecting with the egg; still, every little drop in sperm quantity or quality potentially reduces the chance of conceiving a pregnancy. As the song "Every Sperm Is Sacred" from *Monty Python's the Meaning of Life* goes: "Every sperm is sacred. Every sperm is great . . ."

A Cluster of Unfortunate Events

A hidden player in the men's infertility picture that often goes unrecognized: low testosterone. As previously mentioned, testosterone levels have been declining—by 1 percent per year since 1982, according to research from the United States and several European countries. In the male fertility-

---

* Men's reproductive function decreases with age, too, in ways that compromise fertility. As men age, they naturally experience a decrease in testosterone levels and sperm counts, as well as more erectile dysfunction and ejaculatory dysfunction, all of which can make it harder for a man to do his part in conceiving a pregnancy.

foiling equation, this makes sense, since adequate testosterone is needed to produce healthy sperm, and many of the factors that can lower sperm count can affect male hormone levels, too. They're parallel manifestations of a common source of disruption.

Given this testosterone decline, it's not surprising that the use of testosterone replacement therapy has increased fourfold among men between the ages of eighteen and forty-five and threefold among older men in the past ten years. After all, many men are aware that low testosterone levels can set the stage for muscle loss, increased abdominal fat, weakened bones, and memory, mood, and energy problems, symptoms many men desperately want to avoid; however, many don't realize that low testosterone often correlates with a lower sperm count. Here's the surprising, counterintuitive fact of life: testosterone replacement therapy comes with its own downsides, including . . . wait for it . . . lowered sperm count!

Here's how this happens. When a man wears a testosterone patch or applies a testosterone gel, the hormone enters his bloodstream and his testosterone levels go up. Sounds good so far, right? But his brain interprets this rise as a sign that there's plenty of testosterone, so it sends signals to the testes to stop making more; this in turn causes a decline in sperm production. The result can lead to something of a vicious cycle, in which men with low testosterone and low sperm quality opt for testosterone treatment and end up with even lower sperm quality. In fact, testosterone replacement therapy has been studied as a method of birth control because 90 percent of men can have their sperm counts drop to zero while they're on it.

When Bad Habits Go Up, Guess What Doesn't?

Adding to these sexual frustrations, increasing numbers of younger men are grappling with a problem that's long been thought to be an older man's affliction: erectile dysfunction (ED). Believe it or not,

26 percent of men who present with some degree of ED are now under age forty. In a study that evaluated nearly eight hundred men seeking help for the first time for erectile dysfunction, researchers found that the average age at which men sought medical attention for not being up to the task dropped by seven years between 2005 and 2017.

Whether it's due to unhealthy lifestyle factors such as smoking, heavy drinking or drug use, higher rates of anxiety, or an increase in porn consumption (which can deplete dopamine reserves due to overstimulation), the result can be the same: trouble getting or keeping an erection during real-life sexual intercourse. In addition, preliminary evidence suggests that exposure to certain environmental agents, such as pesticides and solvents, as well as arsenic in well water, can compromise erectile function. *Add these to the list of sexual hazards in the modern world!*

Hard Truths, Painful Emotions

Despite the fact that the decline in sperm counts presents a formidable threat to men and couples alike, there's often a reluctance to accept this reality, even when men and women are aware of it. In other words, there's often a disconnect between knowing a problem exists and being willing to accept it. For example, research suggests that in many countries "male infertility remains a hidden, highly stigmatized problem—laden with feelings of inadequacy, and often spoken of, derogatorily, as in *shooting blanks*— and it leads to feelings of emasculation," notes Marcia C. Inhorn, PhD, a professor of anthropology and international affairs at Yale University. This isn't entirely surprising since historically a man's virility has been considered an integral part of his sense of masculinity. But "many people have absolutely no idea that male infertility is something different from male impotency," she adds.

For thirty years, Dr. Inhorn has conducted research on male infertility in the Middle East. In this part of the world, certain genetic sperm defects and male factor infertility problems are common and often run in families. Yet, even when their husbands are discovered to be the infertile ones, women are often blamed for the infertility, and sometimes women try to help their infertile husbands save face by claiming the infertility problem as their own, Dr. Inhorn notes. "It's often done out of love. They do it because they don't want their male partners to be humiliated."

Granted, it's often hard for men to come to terms with the reality that they aren't as virile as they presumed they were, even when they're presented with evidence that this is the case. In one study, researchers from the UK asked men experiencing infertility to share their thoughts and feelings about what they were going through. All characterized their desire to procreate as "a taken-for-granted expectation" and "part of being human," so merely seeking help for fertility issues was viewed as a sign of "weakness" and caused them shame and embarrassment. After being diagnosed with infertility, subfertility, or having defective sperm, the men said such things as "You almost feel as if you're not a man. You cannot do the biological thing" and "Part of being a man is being able to produce children.... When they tell you that you can't, that your semen's no good, it's like ... taking a bit of masculinity away from you." Or, "I know it's my fault and it's my problem and my partner could have kids with somebody else.... She's got the option. Whereas I haven't got the option to do that."

Sharon Covington, MSW, has spent thirty-five years in the field of reproductive mental health, providing specialized fertility counseling to individuals and couples in the greater Washington, DC, area. Editor of the book *Fertility Counseling: Clinical Guide and Case Studies*, Covington is also director of psychological support services at Shady Grove Fertility, the largest fertility practice in the United States with thirty-two centers throughout the country, and she routinely counsels men and women who experience emotional stress from their fertility challenges. While this type of news is difficult for either gender to accept, "it comes as a real shock when a man finds out he has a low sperm count or other male factor fertility problems," Covington says. The surprise factor stems in part from the fact that men don't have regular wellness visits to check their repro-ductive function or prenatal fertility checks; only when they have trouble getting their female partner pregnant do men find out they may have a fertility problem.*

Often, women who are faced with fertility challenges seek immediate support, whereas men are more likely to keep the disappointing news to themselves. "Among men, it's not the kind of thing they'd ever share in a locker room or with a buddy over beer," Covington says. "It becomes a very private, isolating experience." Not surprisingly men's lack of open-ness about their infertility is a risk factor for experiencing depressive symptoms. Nor does it help that men with fertility problems have a significantly lower-quality sex life compared to male partners who don't have problems with fertility, as one study found.

When researchers from Montreal examined the content of online discussion boards for men with fertility problems, they discovered that various types of social support—including emotional and informational support—were both sought and provided by those writing on these boards. When the cause of childlessness was male factor infertility, men wrote such things as "I'm really disappointed [and] I have a feeling [my wife] holds me responsible for it." One guy wrote, "What I hate most are the thoughts I can't help about what people think when they talk to me. Is it pity? . . . I'm so conflicted because I know I'd feel the same way as those people if the tables were turned."

Hazards of Playing the Waiting Game

Complicating the rising challenges to male fertility, many couples in West-ern countries are now waiting until their thirties to start a family. So they may not discover that one or both of them have fertility problems until they have only a narrow window of opportunity to take advantage of assisted

---

* As Cynthia Daniels, PhD, a professor of political science at Rutgers University, noted in her book *Exposing Men*, "Politically, the need to reinforce the myth of male in-vulnerability has resulted in a lack of attention to questions of male reproductive health." Clearly, this does men a grave disservice in the grand scheme of things.

reproductive technology (ART) such as in vitro fertilization (IVF). Since there aren't any treatments for improving sperm production in subfertile men, the only effective option is for the couple to undergo ART, which is not only expensive, but also invasive for the woman.* Ready for a shocker? Male factor infertility is the only medical situation that's treated by administering a painful procedure to a woman because of a problem that afflicts her male partner.

Another potential glitch: a compelling body of research shows that as men age, particularly as they reach the north side of forty, their sperm is more susceptible to mutation, which can increase the risk that their children will be born with disorders such as autism and schizophrenia or Down syndrome. A man's age also can affect his female partner's miscarriage risk. Studies suggest that for men ages forty and older, their partner has a 60 percent increased risk of experiencing miscarriage, compared to fathers under thirty; the risk appears to be stronger for first-trimester pregnancy losses, which are more likely to be chromosomally abnormal. That's right—a pregnant woman is more likely to miscarry when her partner's sperm is faulty, but neither partner may realize this.

Sadly, there's no easy solution to the problem of aging sperm when it comes to the prospect of achieving and sustaining a pregnancy. Assisted reproductive technology may seem close, but it's not a panacea.†

In recent years, fears about declining sperm counts—and concerns about the lack of preventive screening for male factor infertility—have spawned the development of several at-home sperm tests that allow a man to collect a semen sample, place it in a special sperm-spinning device, and get a reading of his sperm count, right in the privacy of his own home.

---

* As a 2018 article in *Prospect* magazine suggested, a "technofix" solution may be on the horizon: "There may come a day when even a complete inability to produce viable sperm in the testes might not be an obstacle to a man having a biologically-related child. In 2016, biologists at Kyoto University reported that they had created 'artificial sperm' from the skin cells of adult mice by reprogramming them."

† For one thing, children conceived through ART, especially ICSI, have a higher risk of autism spectrum disorder and intellectual impairment.

But because they're so new, the accuracy and reliability of these home sperm-count tests have yet to be determined—and they don't assess other factors, such as motility or morphology. Meanwhile, sperm cryobanking services, such as Legacy, are now making it possible for younger men to bank their potent sperm for the future in case they want to have children down the road, just as egg-freezing services allow women to do their part.

Contrary to public perception, fertility challenges are an equal opportunity problem between the sexes, not just a woman's issue. And the declines in sperm count and quality that are occurring in the modern world aren't helping matters. It really does take two to tango or fox-trot—or produce a viable pregnancy and healthy offspring. The difference is, just because a man doesn't hear his biological clock ticking doesn't mean it isn't marking time.

3

IT TAKES TWO TO TANGO:
Her Side of the Story

Reproductive Wrongs

When Margaret Atwood's novel *The Handmaid's Tale* was first published in 1985, people responded primarily to its disturbing depiction of women living in what might be described as a feminist's nightmare: a world in which women are under strict patriarchal and social control, forbidden from having jobs or money of their own and assigned to various classes—from chaste, childless wives to housekeepers to reproductive handmaids whose purpose is to become impregnated by the men whose homes they inhabit so they can then hand over their babies to the men's "morally fit" wives. At the time, no one thought the portrayal of catastrophic declines in birth rates could be linked to toxic chemicals in the air and water; somehow, *that* seemed like dramatic license on the author's part. But now the novel and the series it engendered seem disturbingly prophetic.

Along with the precipitous drop that has occurred in sperm counts and fertility rates in Western countries, the rate of gestational surrogacy, a consensual version of the scenario described in *The Handmaid's Tale*, has steadily been increasing—about 1 percent per year between 1999 and 2013. This trend reflects a downturn in fertility. While the dramatic decline in sperm counts is an important factor in the fertility slump that's being seen in many parts of the world, changes in women's reproductive function are also occurring, and many have links to the same lifestyle and environmental culprits that are affecting men's reproductive status.

Before I get to those, some facts are in order to illustrate the big picture. Worldwide fertility dropped by 50 percent between 1960 and 2015, and in some countries the decline has been even steeper. For example, between 1901 and 2014 the total fertility rate in Denmark dropped from 4.1 children per woman to 1.8 children per woman. At first glance, it's easy to attribute the decline to social trends such as women choosing to have their first pregnancy at older ages and couples' desire for smaller families. Those things undoubtedly contribute to the shift. But it's not that simple, because fertility declined *at all ages* during this same time period. And, surprisingly, the decline in the ability to conceive a pregnancy and carry it to term—what's called impaired fecundity—was actually more dramatic in younger women.* And here's the real shocker: in the first decade of the twentieth century, women over age thirty in Denmark had higher fertility rates than women under thirty had between 1949 and 2014. Looked at another way, the average twentysomething Danish woman today is less fertile than her grandmother was at thirty-five. *No bueno!*

The picture is almost as bleak in the United States, with total births per woman dropping by more than 50 percent between 1960 and 2016. It isn't clear how much of this baby scarcity stems from economic, educational, sociological, or environmental factors, but this much is undeniable: in 2017, the total birth rate for women in the United States was 16 percent below what is considered necessary for our population to replace itself over time. That's obviously cause for concern—this was true in 2017 and it's still true in the time of COVID-19. To borrow a phrase from William Shake-

---

* When a colleague and I looked at the change in impaired fecundity by women's ages from 1982 to 1995, we were surprised to see that those between the ages of fourteen and twenty-four, the youngest women, had experienced a 42 percent increase in impaired fecundity, while women ages thirty-five to forty-four experienced only a 6 percent increase. This suggests that something besides aging and delayed childbearing is affecting fecundity.

speare, these trends suggest that something is rotten (or at least trouble-some) in the state of Denmark, the United States, and elsewhere.

Indeed, there's compelling evidence that diminished ovarian reserve (DOR)—a condition in which the number and quality of a woman's eggs is lower than expected for her biological age—is occurring more frequently than in previous generations, and that the risk of miscarriage (pregnancy loss before twenty weeks' gestation) has been rising among women of all ages.

While the recent increase in reproductive challenges among women may not be quite as dramatic as those in men, we may not be getting the full story of what's going on. For one thing, there are more studies on men's reproductive functionality, partly because, *well*, more medical studies are conducted on men—period. (Yes, there's a gender gap when it comes to medical research, as well as in pay equity, employment opportunities, dry-cleaning fees, and other elements in our modern world.) As far as research on reproductive health goes, there may be an element of practi-cality at work here: men's genitals are on full display, and a sperm sample can be obtained from an ejaculation that's provided by a man without too much effort or trouble.

With women, by contrast, no fluid offering can reveal her reproductive potential or limitations. In women, the inner workings of reproductive capac-ity are more complicated and are hidden from view. For example, there's no easy way to count the number of eggs a woman has in reserve.* And even if she has plenty of eggs remaining and she ovulates regularly, a woman has no way of knowing if her fallopian tubes are blocked, if her uterus is hospita-ble to a fertilized egg, or if the right hormones will be released in the right amounts at the right times to provide a safe haven for an embryo—until she tries to get pregnant. So gauging a seemingly healthy woman's baby-making prospects is a trickier proposition than gauging a man's.

---

*   To guesstimate ovarian reserve, doctors often measure blood levels of follicle-stimulating hormone (FSH), estradiol, inhibin B, or anti-Mullerian hormone (AMH)—but these aren't considered reliable indicators. Which means these results may provide false hope or instill unnecessary worry.

Don't Know Much About Biology

Despite the fact that the female body is a baby's first home, many women know less than one might expect about the ins and outs of reproductive health. This isn't just an educational problem; it has real, practical implications for reproductive success. Study after study has found that fertility awareness among women is shockingly low. On average, women answer correctly 50 percent of questions about the causes and prevalence of infertility; medical students fare only slightly better, typically earning a D instead of an F. In one study involving a thousand women between the ages of eighteen and forty in the United States, 40 percent of participants expressed concerns about their ability to conceive, but one-third were unaware of the adverse effects that sexually transmitted infections, obesity, or irregular periods could have on their ability to procreate. Even more startling: 40 percent were unfamiliar with the ovulatory phase of their menstrual cycles, which is the only time fertilization can occur.

Given the confusion about ovulation, here's a brief refresher: Ovulation occurs around day fourteen in a twenty-eight-day menstrual cycle (day one is the first day of a woman's period), when a surge in luteinizing hormone (LH) causes one of a woman's ovaries to release a mature egg. (Though the average cycle is twenty-eight days long, anything between twenty-one and forty-five days is considered normal, and normal periods last two to eight days.) To identify when she's on the verge of ovulating, a woman can track several things. First, changes in her cervical mucus: it becomes thin, clear, and slippery like egg white right before ovulation. Or she can monitor her basal body temperature—first thing in the morning, before getting out of bed—because it will rise about half a degree when ovulation occurs. Or she can use an ovulation predictor kit, which forecasts ovulation twelve to twenty-four hours in advance, after she pees on a stick. (Note that these techniques are far from foolproof as contraceptive methods; they're more useful for a couple trying to conceive.)

After the egg is released, it slowly travels down the closest fallopian tube toward the uterus, whose lining has been prepared for the possibility of pregnancy, thanks to increasing levels of the hormone progesterone. If healthy sperm have swum upstream from the vagina through the cervix and into the fallopian tube, one can complete its mission and fertilize the egg there. (Amazingly, after sexual intercourse, sperm can stay alive in a woman's reproductive tract for at least five days, especially if they're protected by fertile cervical mucus. Which means a couple doesn't need to have unprotected sex on the exact day that a woman ovulates in order to get pregnant; there's a window of opportunity of approximately three days leading up to ovulation.) Once it's fertilized, the egg travels into the uterus, and if everything goes right, implantation will occur in the lining of the uterus; if it doesn't, the unfertilized egg will pass out of the woman's body during her period.

Those are the basic facts of a woman's reproductive function—and they haven't changed with the passage of time. But recent decades have seen some baffling shifts in female reproductive development, health, and fertility. Among others, there has been a downturn in sex drive among men and women of all ages, as previously mentioned. Low sexual desire is the most common sexual problem among women at midlife, affecting 69 percent of women over age forty, according to one study. In an unfortunate double whammy, among postmenopausal women, low libido is often tied to erectile dysfunction in their male partner. Whether these sex drive nosedives stem from stress, medication use, chemical exposures,* or other factors, there's no denying that they're a bummer in the bedroom.

---

* Interestingly, one study I was involved in found that premenopausal women who had the highest urinary concentrations of a di-2-ethylhexyl phthalate (DEHP) metabolite (from exposure to a chemical plasticizer) were two and a half times more likely to report that they always or often lacked interest in sexual activity. This may be because phthalates such as DEHP are well known for having antiandrogenic effects, such as lowering testosterone, which plays a key role in sex drive for men and women; they also may interfere with estrogen production in women, thus suppressing women's libido.

An Accelerated Timetable

In an unanticipated turn of events, in some parts of the world, including the United States, girls are maturing earlier and experiencing what's called early puberty; that is to say, they experience earlier breast development and start of menstruation, sometimes before age eight. The alarm first sounded on this issue back in 1997 when a study showed that by age seven, 27 percent of African American girls and 7 percent of white girls were showing signs of breast and/or pubic hair development. The researchers found that on average, African American girls were beginning puberty between the ages of eight and nine and white girls by age ten—six months to a year earlier than girls in previous studies.

In 2006, girls in Denmark had developed glandular breast tissue (the hallmark sign of puberty) a full year earlier than girls born in the same region in 1991. Similarly, the age at which girls began getting their first period also decreased; in the Danish study, it was three and a half months earlier in girls than in their mothers. In Japan, the onset of menstruation shifted from 13.8 years for girls born in the 1930s to 12.8 years for those born in the 1950s to 12.2 years for those born in the 1970s and 1980s.*

These may not sound like dramatic differences, but to the girls experiencing them, they're significant. Not many girls in elementary school are thrilled by the prospect of having to carry tampons or menstrual pads in their cartoon-themed backpacks. Girls who go through early puberty may have mood swings before their peers do, and that can lead to social isolation, depressive symptoms, and use of illicit substances such as alcohol or recreational drugs. And because these girls often look older than they are, sexual attention may be directed toward them before

---

*   Trends in women's reproductive landmarks continue to shift around the world. A meta-study involving more than half a million women from ten countries found that women born between 1970 and 1984 started their periods a full year earlier than those born before 1930. Another noteworthy change: the prevalence of never giving birth (called nulliparity) rose from 14 percent among women born between 1940 and 1949 to 22 percent among those born between 1970 and 1984.

they're emotionally ready to handle it. All of which can lead to a premature loss of innocence.

The extent to which these precocious shifts bother girls varies considerably, but being ahead of the pubertal curve is often uncomfortable, as Kate remembers all too well. After developing breasts at age nine and getting her first period at ten, Kate experienced relentless teasing from boys at school, who often called her "Brenda Starr" or a "brick house," referring to her well-endowed, voluptuous figure. "I definitely got more attention from boys, some of it was appreciated because I was fairly boy crazy, but some of it was not, especially the pinching and name-calling," recalls Kate, now forty-five, whose daughter also went through puberty early. "For me, the worst part was that I gained ten pounds in a summer when I was ten and my mood swings were off the charts." The only upside, as far as Kate was concerned, was that she became something of a menstrual mentor to friends who got their periods and started wearing bras after she did and sought her advice.

As challenging as early puberty can seem while a girl is going through it, there are often enduring ripple effects, such as higher levels of psychological distress and body-image problems as an adult. There are also potential long-term implications for a woman's physical health. Most notably, an earlier age at the onset of menstruation has been linked to an increased risk of breast and endometrial cancers because the risk of those cancers increases with the number of menstrual cycles a woman has throughout her life.

Female Trouble in the Fertility Department

Other worrisome shifts are occurring in the reproductive realm for women. After spending years or even decades trying *not* to get pregnant, a woman who wants to have a baby might assume she'll get pregnant quickly with well-timed, unprotected sex. But it doesn't work out that way for everyone, especially these days. The truth is, human reproduction is highly inefficient, especially compared to that of the majority of mammalian species. During a given menstrual cycle, people have—at best, depending on their age—an approximately 30 percent probability of conception with well-timed, unprotected sex.*

To be fertile, a woman needs to have functioning ovaries, a reserve of healthy eggs, healthy fallopian tubes, and a healthy uterus. Any medical condition that affects these organs can contribute to female infertility. One such condition is polycystic ovary syndrome (PCOS), a hormonal and metabolic disorder. It's characterized by irregular periods, excess facial or body hair, acne, weight gain, and multiple cysts on the ovaries; fallopian tube obstructions or scarring also can occur with PCOS. Another medical problem that impacts fertility is endometriosis, an often painful disorder in which tissue that normally lines the uterus gets displaced and grows outside the uterus. Fibroids, benign growths of muscle and fibrous tissue that develop in the uterus, can also decrease the chance that a woman conceives. And there are signs that all these reproductive disorders are increasing.

For example, a retrospective study of nearly seven thousand women in Canada found more than a threefold increase between 1996 and 2008 in the number of women between the ages of eighteen and twenty-four who were newly diagnosed with endometriosis. At first blush, it's difficult to tell whether this diagnostic swell is because these conditions are occurring more frequently or if doctors have simply become better at recognizing the symptoms and making the correct call. I suspect it's a bit of both.

Astonishingly, Isabel, thirty-two, a school social worker in New York City, didn't find out until she'd been trying to get pregnant for a year, without success, that she had Stage IV endometriosis, the most severe form. Not until she had a CT scan and exploratory surgery to investigate why she was having trouble getting pregnant was her endometriosis discovered. During the procedure, surgeons removed as much of the misplaced endometrial tissue as they could find and removed her damaged

---

* By contrast, rodents have a 95 percent probability of getting pregnant and rabbits have a 96 percent chance. Lucky them!

fallopian tubes. After that, Isabel was able to get pregnant through IVF and now has a two-year-old son.

She continues to wonder how or why she got endometriosis, because no one else in her family has had it. "I work with ten other women who've recently gone through fertility treatment, and we often talk about what's going on in our environment that's causing so many fertility problems," she says. "Maybe there's something in the water or in our food. *Who knows?!* In this day and age, I feel like nothing is healthy anymore."

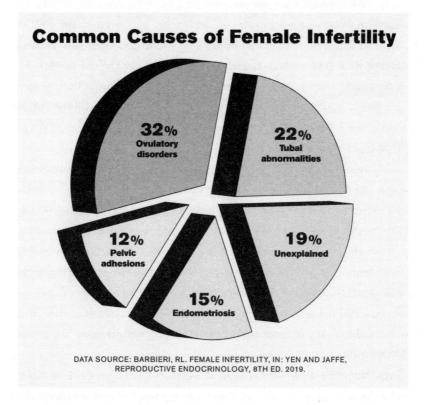

Common Causes of Female Infertility

32%
Ovulatory disorders

22%
Tubal abnormalities

12%
Pelvic adhesions

19%
Unexplained

15%
Endometriosis

DATA SOURCE: BARBIERI, RL. FEMALE INFERTILITY, IN: YEN AND JAFFE, REPRODUCTIVE ENDOCRINOLOGY, 8TH ED. 2019.

Dashed Eggspectations

Of all the potential fertility foilers, ovulation disorders are responsible for the largest proportion of female causes of infertility, with advancing age playing a primary role. Amazingly enough, a female is born with all the eggs she will ever have—approximately 1 million to 2 million, which is far more than she will ever need. By the time she reaches puberty, approximately three hundred thousand eggs will remain, all but one of which are quiescent and idle during any given month. (Usually, only one egg is released during ovulation, but some fertility drugs stimulate the ovary to release more than one egg, which is why fertility drugs often lead to multiple births.)

As the decades pass, the number of eggs a woman has in storage dwindles steadily to an average of twenty-five thousand at age thirty-seven, then hits an even more dramatic slide, down to one thousand at age fifty-one (the average age of menopause in the United States). As with sperm, it's not just a numbers game, however. In addition to this age-related decline in the *quantity* of eggs a woman has, there's also a substantial decrease in the *quality* of healthy, viable eggs in her body as she approaches forty.

It has always been difficult for women to get pregnant as they get older. But this didn't used to be as much of a problem because women had babies at younger ages. Now they are increasingly delaying having children. And though that may well be a good thing from a social perspective, it isn't from a reproductive one. It's ironic that when it's biologically easiest for a woman to get pregnant and give birth, many women aren't yet thinking about having a child. Unfortunately, Mother Nature hasn't kept up with women's shifting desires in the baby-making department and extended our reproductive life spans accordingly.

Admittedly, there are substantial variations in the rate at which a woman's eggs die off or sustain their quality, based on genetic, environmental, and lifestyle factors. It's not just a linear effect as her birthdays pass. That's always been true, but new actors on the environmental stage and in the lifestyle arena may be impacting these rates. I'll get to that soon.

First, it's interesting to note that while the average age of menopause is *not* decreasing, there's evidence that diminished ovarian reserve (DOR) is occurring more frequently than in previous generations. The prevalence of DOR increased from 19 percent to 26 percent between 2004 and 2011 among women seeking assisted reproductive technology (ART) treatment in the United States; that's a 37 percent increase in just seven years. While it is possible for a woman with DOR to conceive naturally, it's much more difficult, and many women don't discover they have diminished ovarian reserve until they have difficulty conceiving.

Sometimes this kind of trouble feels as if it comes out of left field. For example, by the time she turned thirty-one, Elissa, a trim lawyer who often runs 10Ks, had given birth to two healthy boys three years apart. When she was thirty-four, she and her husband decided they wanted to have a third child and expected it to happen as easily as it had with the first two. It didn't. After trying to get pregnant for nine months, Elissa went for a fertility evaluation and was told that she had "old eggs": in simple terms, her eggs had aged prematurely, and the quality of her remaining eggs was relatively poor, considering her biological age. Realizing she was fortunate to have two children already, Elissa tried to joke about her "rotten eggs," but she says, "Inside, I felt broken." She couldn't understand why this had happened to her.

To improve their chances, the couple opted to undergo in vitro fertilization, and after two unsuccessful IVF cycles, the third IVF cycle resulted in Elissa's getting pregnant. Unfortunately, when Elissa had a miscarriage eleven weeks into the pregnancy, she began wondering what she might have done to jeopardize the pregnancy.

That's not uncommon, says Alice Domar, PhD, chief psychologist at Boston IVF at Beth Israel Deaconess Medical Center and author of *Finding Calm for the Expectant Mom*. After a miscarriage, it's not unusual for women to retrace their recent histories to try to pinpoint what went wrong. "People need to find a reason; it's hard for them to have something terrible happen to them randomly," Dr. Domar says. But miscarriages rarely occur because of something a woman did.* More often than not, miscarriages are tied to chromosomal abnormalities.

When Time Is the Enemy

The truth is, age is not on a woman's side when it comes to having or sustaining a healthy pregnancy. As women get older, they tend to experience a negative triple whammy, an increase in the risks of three interrelated adverse reproductive outcomes—infertility, miscarriage, and chromosomal abnormalities (including trisomy 21, which is the presence of three copies of chromosome 21, also known as Down syndrome). To put this in perspective, consider this: between the ages of twenty-five and thirty-five, women have a 25 to 30 percent chance of getting pregnant with well-timed, unprotected sex in any given month, a 10 percent risk of having a miscarriage, and a 1-in-900 chance of having a baby with Down syndrome; by contrast, women who are forty have a 10 percent chance of getting pregnant with well-timed, unprotected sex in any given month, a 40 percent miscarriage rate, and a 1-in-100 chance of having a baby with Down syndrome. The odds are not in their favor, on multiple fronts.

Because the early weeks of pregnancy are when 80 percent of all miscarriages occur, some women heed the twelve-week rule and wait until

---

* Research has found that 50 to 66 percent of pregnancies that ended in miscarriage were chromosomally abnormal. The rate of chromosomal abnormalities in earlier pregnancy losses, before the woman even knows she's expecting, is probably even higher. As you may recall from high school biology, chromosomes are the gene-containing structures inside the nucleus of each cell. In humans, each cell normally contains twenty-three pairs of chromosomes. During fertilization, when an egg and sperm fuse, the two sets of chromosomes (from a man and a woman) come together. If a fertilized egg has an abnormal number of chromosomes—or if it has duplicated, missing, or incomplete ones—problems with implantation of the embryo or early miscarriage can occur.

the second trimester to go public with their news; at that point, they're not completely out of the woods, but the risk of pregnancy loss declines as women enter the second trimester. The exception: the entire risk profile continues to climb along with a woman's age.

Playing the Odds

As women age, it becomes harder to conceive and deliver a healthy baby.

Women's Age	Pregnancy Rate Per Month	Miscarriage Rate	Down Syndrome Risk
25–35	25–30%	10%	1/900
35	20%	25%	1/300
37	15%	30%	1/200
40	10%	40%	1/100
45	5%	50%	1/50
50	1%	60%	1/10

DATA SOURCE: HTTP://MARINFERTILITYCENTER.COM/NEW-GETTING-STARTED/INFERTILITY-BASICS

A substantial proportion of women's perceived infertility as they get older is the result of undetected pregnancy loss—meaning, a woman loses the embryo before she even realizes she's pregnant. These early losses are largely due to chromosomal abnormalities—which are contributed by the man, the woman, or both partners—in the fertilized egg. The only way for a woman to find out that she's pregnant is to test her urine for elevated levels of human chorionic gonadotropin (hCG) hormone, which can't be detected in urine until six or seven days after conception; however, many women wait until they have missed a period to do a pregnancy test, by which point they may have already lost the pregnancy, especially if they've reached the north side of forty. Perhaps this is one reason why Dr. Juan Balasch, an obstetrician-gynecologist at the University of Barcelona in Spain, has suggested that female fertility "has a 'best-before date' of 35," while the fertility shelf life for men extends to age forty-five to fifty (and sometimes beyond).

Miscarriage Mysteries

Even when women of any age do succeed in getting pregnant, their pregnancies seem to be increasingly threatened these days. In recent years, the rate of miscarriages has been on the upswing among women in the United States, regardless of the expectant mother's age. From 1990 to 2011, the risk of miscarriage increased by 1 percent per year among pregnant women in the United States, according to a 2018 study by the Centers for Disease Control and Prevention. It's worth noting that this is the same rate at which sperm count and overall fertility have declined in Western countries. All of these fertility-related rates are going south at approximately the same pace—the new 1 percent effect is real and worrisome and has nothing to do with income!

Not surprisingly, many women experience depression and/or anxiety after a miscarriage. According to Dr. Domar, "the minute a woman realizes she's pregnant, that's a baby to her—she's thinking about names and a nursery. So if a miscarriage occurs, there's a potential for it to be perceived as a death, and the grieving process can be intense." Offering a refreshing dose of reality, former First Lady Michelle Obama revealed in her memoir, *Becoming*, that after she had a miscarriage that was "lonely, painful, and demoralizing almost on a cellular level," she and Barack relied on IVF to conceive both Malia and Sasha. As she wrote, "Fertility is not something you conquer."

Women who have miscarried often feel betrayed by their body, having been raised with the notion that the female body is conditioned to produce babies. When a woman's doesn't, "there's often a sense of her body being defective in some ways, which can have a profound effect on her self-image, body image, and self-esteem," notes Sharon Covington, MSW, director of psychological support services at Shady Grove Fertility. Even those who are fortunate enough to conceive again may be more vulnerable to depression in the month after they give birth to a healthy baby. For those who experience recurrent miscarriages the emotional effects can be profound and long-lasting. Similarly, ongoing fertility problems can have substantial ripple effects not only on a couple's life together but also on a woman's state of mind and sexual well-being.

After having two miscarriages, Diane, forty, was thrilled when her next pregnancy progressed easily to sixteen weeks. Given her age and elevated risk for having a baby with chromosomal abnormalities, including Down syndrome, Diane scheduled an appointment for amniocentesis, a prenatal procedure in which a small amount of amniotic fluid is drawn from the uterus to test for chromosomal conditions and fetal infections. (Amniocentesis is routinely done in pregnant women over thirty-five.) The doctor performing the procedure had trouble extracting amniotic fluid because Diane's placenta was placed on the front wall of her womb, and he had to reinsert the needle several times to get a proper fluid sample. Diane left the appointment feeling shaken—until she received the news that she was carrying a healthy baby girl. Diane and her husband named her Ella Rose and imagined holding her in a cozy onesie; Diane's two children from a previous marriage did, too.

At her next prenatal checkup, her ob-gyn couldn't find the baby's heartbeat. A subsequent ultrasound confirmed the devastating news that Ella Rose had died in the womb. Diane had to wait until her body was ready to deliver her deceased baby naturally because of the substantial risk of excessive bleeding if doctors induced labor. "It was the longest three weeks of my life," she recalls. Whether the miscarriage was due to Diane's age or the amniocentesis—the procedure carries a .1 to .3 percent risk of miscarriage—couldn't be determined, but it was deeply upsetting. "I worried that if I couldn't give him a child, it would put my marriage in jeopardy because he really wanted one," Diane says. "I felt like I was inadequate."

What she didn't know until many years later is that the problem causing her recurrent miscarriages, defined as three or more consecutive pregnancy losses before twenty weeks' gestation, could have been her husband's, rather than hers.

In fact, recent research found that in couples who experience recurrent miscarriages, the men have twice the level of DNA fragmentation in their sperm and four times higher levels of reactive oxygen species in their semen, which can cause DNA damage to sperm, than men whose partners didn't have a history of repeated miscarriages. In couples with recurrent pregnancy loss, the men also had reduced sperm motility and morphology, compared to their peers. As the semen quality goes down, the risk of miscarriages goes up because of bad sperm, which, as you've read, are increasingly common. And more often than not, it's the woman who bears the brunt of the emotional distress because she's the one carrying the embryo, and the man's role in miscarriages is not acknowledged. It's common practice for women who have had recurrent miscarriages to be sent for reproductive assessments to try to figure out why; the latest findings suggest that their male partners should get checked out, too.*

Some data also suggest that recurrent pregnancy loss may be on the rise. Between 2003 and 2012, the incidence of recurrent miscarriages increased by 74 percent among a cohort of 6,852 women ages eighteen to forty-two in Sweden. That's a rapid increase in the span of just nine years! Which is why the researchers speculated that it might be due, at least in part, to environmental factors, though they didn't hazard a guess as to which ones are to blame.

False Hope from Famous Baby Bumps

The media often report on celebrity moms who have kids while in their forties (think Rachel Weisz, Janet Jackson, Nicole Kidman, and Halle Berry)—and the celebrities act as if it's no big deal, just another blissful day in Hollywood. That's great for them, but it's potentially misleading for ordinary Janes because we seldom hear whether the celebrities had any help in the fertility department. Some famous women have taken

---

*   In recent years, a theory has emerged that Henry VIII may have been the reason why several of his wives, two of whom he famously executed, suffered repeated miscarriages.

fertility drugs, undergone IVF, or used donor eggs—but the backstory isn't always told. Granted, it isn't the public's business, but these omissions can lead younger women to think that they, too, can put off having kids until their forties.

Women significantly overestimate the chance of pregnancy at all ages. A survey of nearly twenty-one hundred women in the United States and Europe revealed that 83 percent of the US women said they underestimated the amount of time it would take them to get pregnant. Similarly, women of reproductive age know little about the effects of aging on fertility and pregnancy, and many are unfamiliar with the success rates for infertility treatments or the high risk of miscarriage. In a study at Northwestern University, researchers asked three hundred women between the ages of twenty and fifty to estimate the probability of pregnancy through natural conception and with assisted reproductive technology at five different ages (twenty-five, thirty, thirty-five, forty, and forty-five). Age thirty-five was the tipping point where the women's estimates became significantly off base; for example, their estimates of a woman's probability of nonmedically assisted pregnancy at age forty was nearly 50 percent higher than the published research suggests.

As environmental factors and advancing age continue to influence a woman's chances of getting pregnant and carrying the baby to term, it's important for women to be realistic about what's possible in this realm. It can be too heartbreaking to simply roll the dice and hope for a win. Being knowledgeable and having sensible expectations can potentially mitigate some of the reproductive challenges women and men are currently facing. Unfortunately, the onus may be on women to educate themselves about these issues because even obstetrics and gynecology residents are not well versed in age-related fertility issues. They tend to either overestimate the age at which female fertility declines and/or overestimate the likelihood of success using ART. A study of female graduate students at Duke University found that 70 percent believe the media gives the impression that motherhood is possible after forty. Sometimes it is, but sometimes it just isn't.

In recent years, some younger women have become increasingly aware of this discordant reality, which is why elective human egg freezing is on the rise—it allows women more latitude in deferring motherhood. Egg freezing is a bit like having a reproductive insurance policy. But even here age continues to matter: the earlier a woman freezes her eggs, the more effective it is. The ideal window is before age thirty-five, when fertility is still near its peak; but many women don't consider the procedure until they're approaching forty or have even passed that milestone, when the quality of their eggs has already declined. So while there isn't exactly a race to reproduce, there is a time limit on a woman's opportunity to do so or to put her eggs on ice.

Regardless of the reason for it, more women have been using assisted reproductive technology in recent years. From 2000 to 2010, there was a nearly 80 percent increase in egg donation for IVF at fertility centers throughout the United States—from 10,801 to 18,306 per year. In 2017, the ART market throughout the world was estimated to be worth $21 billion (in US currency), and it's expected to increase 10 percent annually until 2025. In recent decades, there's even been a trend called the "graying" of infertility services, whereby increasing numbers of women over forty are pursuing IVF with the hope that it will help them win their own baby-making challenge. But technology can't solve every woman's fertility problem. As women get older, ART cycles (involving fresh embryos from fresh nondonor eggs) that progress to pregnancy are less likely to result in the birth of a live baby because the percentage of pregnancies that end with miscarriage increases.*

Even with the brave new world of advanced fertility treatments—including the alphabet soup of ART, IVF, IUI, ICSI, and others—there may be a point at which even science can't compensate for scrambled eggs or damaged fallopian tubes caused by unhealthy lifestyle practices or

---

* In addition, the chances of pregnancy complications such as fetal growth restriction, hypertension, and premature birth also rise with an expectant mother's age. And children born to older couples are at higher risk of neurodevelopmental problems such as schizophrenia and autism spectrum disorder.

increasingly present environmental hormonal hijackers or advancing age. What doesn't change with age or lifestyle: the expense and discomfort of fertility treatments.

Admittedly, women's reproductive potential is not in as dire straits as depicted in *The Handmaid's Tale*. Not yet, anyway. But the rising prevalences of early puberty, endometriosis, PCOS, miscarriages, and diminished ovarian reserve are certainly troublesome and possibly ominous for the future. The mounting links between a woman's reproductive health and her overall health risks have even spawned a movement to consider fertility status as the sixth vital sign: After all, premature egg loss and early menopause have been tied to an increased risk of developing cardiovascular disease in the future. Strong links have been found between PCOS and an increased risk of getting diabetes and cardiovascular disease. A history of anovulation is associated with an increased risk for uterine cancer, while endometriosis and tubal factor infertility have become red flags for an elevated risk for ovarian cancer. These reproductive disorders, all of which appear to be on the rise, have become forecasts for stormy health problems in the future.

4

Gender Fluidity:
Beyond Male and Female

As the renowned biologist and sex researcher Alfred C. Kinsey wrote in 1948, "The living world is a continuum in each and every one of its aspects. The sooner we learn this concerning human sexual behavior, the sooner we shall reach a sound understanding of the realities of sex." Truer words were never written, but the realities of sexual behavior, gender expression, and gender identity are becoming increasingly complex.

Scientific questions about what makes someone male, female, or non-binary, or straight, gay, bisexual, or asexual, are complex, fraught, and fascinating—and not easy to answer. People have long wondered whether gender identity and sexual orientation are genetically determined or environmentally influenced—whether they're a matter of nature or nurture. In therapy, "gay patients almost always have questions about why they're gay," notes Jack Drescher, MD, a clinical professor of psychiatry at Columbia University who served on the American Psychiatric Association's DSM-5 Workgroup on Sexual and Gender Identity Disorders. "Heterosexual patients don't come in with questions about why they're heterosexual."

The question of whether there's a "gay gene" has been hotly debated for decades. The answer is, it's not that simple. As Siddhartha Mukherjee,

MD, writes in *The Gene: An Intimate History*, "After nearly a decade of intensive hunting, what geneticists have found is not a 'gay gene' but a few 'gay locations' [in a chromosomal region]. . . . The 'gay gene' might not even be a gene, at least not in the traditional sense. It might be a stretch of DNA that regulates a gene that sits near or influences a gene quite far from it." In other words, it's complicated. But that doesn't mean genetic factors don't play a role in influencing sexual orientation; they undoubtedly do.

After its release in 2011, Lady Gaga's song "Born This Way" sky-rocketed to the top of the charts and was quickly embraced by people of various sexualities, partly for its promotion of gay rights and cultural acceptance and partly for its pumping disco-like beat. But some members of LGBTQ (lesbian, gay, bisexual, transgender, and questioning or queer) communities reject the born-this-way description, in large measure because it doesn't necessarily apply to people whose sexuality and/or gender are fluid—a population that continues to grow. According to a 2017 Gallup poll of more than 340,000 adults in the United States, the increase was driven largely by millennials born between 1980 and 1999, 8.1 percent of whom identified as LGBT in 2017, compared to 5.8 percent who did in 2012.

Sexuality Versus Gender Identity

Just as it's increasingly recognized that sexuality resides on a spectrum—meaning that many people aren't exclusively attracted to one sex or the other, that their orientation exists outside binary categories and is sometimes a moving target—the same can be said for gender. To be clear, gender and sex are not the same, though people often conflate the two concepts. A person's sex is determined by biology (based on the presence of certain chromosomes, hormones, and reproductive organs at birth), whereas gender depends on someone's fundamental, inner sense of self, as well as the feelings, behaviors, and attitudes that go along with it. Recently, it has become more widely accepted that with respect to gender identity considerable variations may exist between the poles of male and female. But some experts take issue with the concept of a gender continuum, pointing out that it doesn't allow for the myriad possibilities in establishing one's personal gender. In her book *Gender Born, Gender Made*, Diane Ehrensaft, PhD, prefers to use the term *"gender web*, in which there are intricate, nuanced pathways in three dimensions, side to side, up and down."

Indeed, some transgender people don't experience a consistency of identity in terms of gender, as Jacob Tobia, a gender-nonconforming writer and producer based in Los Angeles, notes in his memoir, *Sissy: A Coming of Gender Story*: "There are many things that I've always known about myself, but my gender just isn't one of them. I didn't know that I was a girl . . . but I wasn't sure that I *wasn't a boy*, either." Tobia has "come to embrace that my gender is more like an onion"—with multiple layers but no distinct core.

In general, gender fluidity reflects the sense that one is a blend or mixture of our cultural notions of masculinity and femininity. The extent of this fluidity can vary from one person to another. "For some, it's the notion that their gender changes over a life course; for others, it changes more frequently, perhaps daily or from hour to hour," explains Ritch Savin-Williams, PhD, a professor emeritus of developmental psychology at Cornell University and author of *Mostly Straight: Sexual Fluidity among Men*. When people report that they wake up feeling one way or another, or that something happens and they suddenly feel more male or female, it isn't clear what sparks that change: Is it something biological, psychological, environmental, or some combination of these influences?

While the perception is that the number of people who identify as gender fluid has increased, it's not clear whether this is true or if it's simply that "people feel greater permission to be gender fluid now because it's a more recognized construct," Dr. Savin-Williams says. These identity issues aren't always easy for people to reconcile, however. With a condition called gender dysphoria, people experience a powerful sense of distress, feeling that their emotional and psychological identity as male or female is out of sync or disconnected from their assigned biological sex. This can begin in early childhood, in which case it's often called early-onset gender dysphoria. For other kids, gender dysphoria can begin around puberty. Some kids who were born female may always have felt that they were born in the wrong gender birthday suit—that they were meant to be boys—whereas others may start to feel this way as they begin to develop breasts and pubic hair and experience other changes associated with puberty.

Gender identity and sexual orientation are often confused with each other, but they are quite different. For some people, their gender identity may change but that doesn't mean the gender they're sexually attracted to shifts, while for others, gender identity and sexual attraction can both fluctuate. Meanwhile, some people who identify as binary—distinctly male or female—could be attracted to the opposite sex or the same sex consistently, or they could be attracted to both sexes (as in bisexual). In a sense, gender identity and sexual orientation are mix-and-match propositions, with a wide array of possible outcomes that can shift over time.

The words that are used to refer to someone's gender are numerous and complex, and the lexicon continues to evolve.* I'm not an expert on this, but I am an expert on how sexual and reproductive development can be affected by environmental influences. Here's what I can tell you about that.

---

*   Gender identity has become so complicated and the potential for social missteps so rife that two professors of sociology and gender studies at the University of California, Los Angeles, recently proposed using "gender-neutral pronouns as the default, with the long-term goal of using *they/them* pronouns for everyone." But some people prefer what are sometimes called neo-pronouns such as *xe/xem* or *ze/hir*. Whether or not you agree with these preferences or suggestions, they illustrate just how much the concept of gender is changing in our world, socially and linguistically. These days, it's safer to ask people what pronouns they prefer to go by or to simply use the person's name, even when referring to that person in the third person ("Julian said . . .").

What Lies Beneath the Gender Blur?

Among the questions contemplated by scientists and mental health experts regarding gender-identity issues: Are changing social attitudes and greater acceptance of people's right to be who they are, deep down inside, influencing the perceived increase? Are biological factors playing a role? Could it be that unseen chemicals in the environment are affecting the development of human sexuality and gender identity?

In a 2019 article in *Psychology Today*, Robert Hedaya, MD, a clinical professor of psychiatry at the Georgetown University School of Medicine, wrote, "It is nothing short of astounding that after hundreds of thousands of years of human history, the fundamental facts of human gender are becoming blurry. There are many reasons for this, but one, which I have not seen discussed as a likely cause, is the influence of endocrine disrupting chemicals (EDCs)."

Many other clinicians and researchers are wondering about this, too. The question of whether chemicals in our midst are affecting gender identity is a bit like the metaphorical elephant in the room—obvious and significant but uncomfortable and difficult to address. One scientific theory suggests that in utero exposure to EDCs, particularly phthalates, which can lower a fetus's exposure to testosterone, may play a role; these chemicals have been associated with an increased risk of autism spectrum disorders (ASDs) in males. Interestingly, ASD and gender dysphoria, two seemingly unrelated conditions, occur together more often than expected. Another theory is that EDCs can interfere with complex biochemical pathways in the brain in ways that may affect how a person associates with his or her physiological sex at birth or expresses their gender through behavior, either of which may result in gender dysphoria.

We also now know that acetaminophen (Tylenol) can have antiandrogenic (e.g., testosterone-lowering) effects. Developmentally speaking, the default brain is female, which means that if an expectant mother is exposed to antiandrogenic chemicals during her pregnancy, her male baby is likely to have a slightly less "male-typical" brain and male-typical behavior as we have shown in our studies. Recently, we found that exposure to hormone-mimicking chemicals during pregnancy can blunt some of the brain-related sex differences that are often seen between boys and girls. Normally, at thirty months of age about twice as many boys as girls are language delayed—meaning they understand fewer than fifty words. When expectant mothers have a low exposure to an antiandrogenic phthalate called dibutyl phthalate (DBP) or they *don't* use Tylenol during pregnancy, the gender difference in language delay in their babies is large; by contrast, when pregnant women *are* exposed to high levels of DBP or Tylenol during pregnancy, there is little difference in language acquisition between boys and girls. Simply put, the language-development difference between the genders becomes blurred with these chemical exposures. I suspect many other qualities do, too.

The truth is, getting to the root of whether EDCs are influencing *gender identity* is difficult. For one thing, we can't rely on animal studies because while many have shown that exposure to environmental chemicals can alter sexual behavior (leading to same-sex mating, for example) and biology (leading to intersex frogs and fish), neither of these outcomes reflects gender identity. With a few exceptions (such as chimpanzees, elephants, and dolphins), most animals are not *self*-conscious, and without a sense of themselves as distinct and separate individuals, gender identity is an irrelevant concept.

Humans are another story because we are self-aware. (Most of us are, anyway.) But with humans it would be nearly impossible, not to mention downright unethical, to perform a randomized, controlled clinical trial in which, say, identical twins, who share nearly the exact same genetic profile, are deliberately exposed to high levels of EDCs during their early years to see what effect this might have on their sexuality and gender identity. Even if it were feasible, the results of such a study wouldn't be informative if the critical period for the development of sexuality and gender identity were during pregnancy, which it likely is since that's when the genitals and brain develop (you'll learn more about this in chapter 5).

Then there's the question of what endpoints should be measured and at what age(s): Should it be based on brain function, social behavior, self-concept, or something else? The answer is further complicated because surveys often rely on binary definitions (male or female), and the issue of gender identity is highly individual.

For these reasons, some researchers are now advocating for the use of scales that measure gradations of femininity and masculinity to assess people's gender identification. When researchers at Stanford University conducted a national survey of more than fifteen hundred adults about their gender identification (based on their self-perception and the way others view them), they found that fewer than one-third of the respondents rated themselves at the maximum of their sex-typical (the sex they were assigned at birth) identification scale. Here's the real eye-opener: for 76 percent of the respondents, their gender profile included overlapping characteristics of femininity and masculinity. When the respondents were given the opportunity to provide open-ended feedback about their responses, it became clear that they considered a range of factors—including their appearance, personality traits, occupation, and hobbies—when indicating their overall sense of masculinity or femininity. For example, a cisgender man—meaning, he was born male and identifies as male—rated himself as a 2 out of 6 on the scale of femininity, and a 5 out of 6 on the masculinity spectrum, explaining, "I consider myself in the metrosexual sort of group. I'm a male who likes females, who is concerned about his skin, clothes, and looks a bit more than most of my friends."

As one of the study authors, Aliya Saperstein, PhD, an associate professor of sociology at Stanford, later wrote in a 2018 State of the Union paper on gender identification, "Gender diversity also exists within the categories of woman and man and within the categories of cisgender and transgender. Much like how differences in political affiliation between Democrats and Republicans are crosscut by ideological positions that range from liberal to conservative, people who identify with the same gender category exhibit variation in their femininity and masculinity—

as self-identified and as perceived by others." In other words, most of us reside somewhere between the poles of extreme masculinity and femininity—and our exact location can vary on any given day.

Between the Gender Lines

The question of what makes someone male or female, beyond the basic anatomical differences, still doesn't have a definitive answer, even biologically speaking. Is it the presence of certain reproductive organs and the absence of others? The presence of secondary sex characteristics such as a deeper voice, more hair, or more muscle mass? Does it have to do with someone's proportion of estrogen and testosterone? While estrogen is typically thought of as a female hormone and testosterone a male hormone, the bodies of both sexes contain these hormones, albeit in different proportions. If a particular woman's body produces more testosterone than those of most females, perhaps because of a genetic anomaly, or if her cells are unusually sensitive to testosterone, she is likely to develop male secondary sex characteristics such as bigger muscles, more facial and body hair, and perhaps an enlarged clitoris.

Over the years, this has been a recurring and thorny issue in elite sports, in particular. Some women who are top competitive athletes naturally have higher levels of testosterone, as well as greater muscle mass, than the average woman does, just as some men have higher levels than others do. But the powers that be in competitive sports have often opted for gender-verification testing. The chromosome test—in which cells were taken from an athlete's mouth, with a cheek swab, and tested for the female-typical XX chromosome pattern—was introduced by the International Olympic Committee during the summer of 1968. The chromosome test was considered a vast improvement over previous sex-verification practices, in which female athletes had to parade naked in front of a panel of physicians and submit to a mandatory genital check or lie on their back with their knees to their chest so the doctors could have a closer look.*

The tests have always been controversial, and some geneticists and endocrinologists weren't fans of the chromosome test because they contended that a person's sex is determined by a confluence of genetic, hormonal, and physiological factors, rather than a single one. It's worth noting that men have never been subjected to such measures to prove or verify their masculinity. But the main point is that considerable variation exists among both men and women when it comes to their anatomy, hormone levels, body composition, and other physiological factors. So one of the underlying concerns with the athletic decisions is, if women who produce extra testosterone naturally are banned from competing in women's athletic events, doesn't that create a slippery slope, potentially opening the doors to prohibiting athletes for other physiological anomalies?†

From multiple vantage points, this is an extremely tangled issue, involving not simply gender identification but also human rights, the right to privacy, the right of people to compete athletically as they were born, and others. After all, elite professional and competitive athletes are

---

* With these tests, "the goal was to prevent men from masquerading as females in women's competition and prevent what was feared could be an 'unfair, male-like' advantage in female athletes born with disorders of sexual development," explains Alison Carlson, a cofounding member of the International Work Group on Sex/Gender Verification Policy in Sports. The issue dates back to the mid-twentieth century when a number of women athletes, many from Eastern Bloc countries, were viewed as hypermuscular or insufficiently feminine in appearance and were blowing away their competition.

† For years, Caster Semenya, a South African middle-distance runner and two-time Olympic gold medalist, has been fighting for the right to compete as a woman and defend her position as one of the world's top female athletes. Legally classified as female at birth, Semenya has seen her gender become the subject of ongoing scrutiny because she has hyperandrogenism: her body produces higher testosterone levels than most women's do. Similarly, India's champion sprinter Dutee Chand was found to have naturally high testosterone levels for a woman after competitors and coaches alerted the International Association of Athletics Federations that her physique seemed suspiciously masculine. In 2014, she was banned for a year from competing as a female and was told she could return to competition if she medically reduced her testosterone level. She refused.

naturally, perhaps genetically, endowed with attributes that give them a competitive edge. Consider the exceptionally long legs of eight-time Olympic gold medalist Jamaican sprinter Usain Bolt, or the incredible wingspan (eighty inches from fingertips to fingertips when his arms are outstretched) of competitive swimmer Michael Phelps, whose twenty-eight medals make him the most successful Olympian of all time. Should people like them be banned from competition because of their natural biological advantages? Should men be disqualified from competition if they have unusually high or low testosterone levels? Where should the gender lines be drawn in competitive sports? These are tricky questions, indeed.

The Ages of Self-Discovery

Anatomy and biology aside, a person's sense of gender identity usually develops in early childhood, often by the age of three. Research has found that babies can distinguish between male and female during their first year of life, but their ability to label and understand gender differences doesn't emerge until sometime between eighteen and twenty-four months. After that, a young child begins to develop concrete associations regarding gender and physical appearances or activities.

An interesting case in point: Several years ago, Tracy's three-year-old son, Aiden, asked her to have a baby so he could have a brother. When baby Barry arrived in 2015, Aiden's wish seemed to come true. But shortly before Barry's third birthday, he took a shine to dressing up in Mommy's clothes, became obsessed with the color pink, and wanted to play with dolls rather than traditional boy toys. One day Barry declared to Tracy, "I'm a girl like Mommy!" Barry had considerable anxiety about his anatomy, and when the two would go to the bathroom together, Barry would ask where Mommy's penis was. "Barry was insistent that I'd lost it and we needed to go find it," recalls Tracy, thirty-four, a graphic designer who works from home. One day, while Tracy was changing him, Barry grabbed his penis and said, *"No penis! No penis!"*—a display of body loathing that was extremely upsetting to his mother.

Shortly after that, Barry insisted on being identified and treated as a girl, dressing only in pink or overtly feminine clothes. Barry's parents rolled with these desires and began referring to Barry as "she," though they haven't changed her name. Even Aiden introduces Barry as his sister. "She is nothing but a little girl—except that everything is completely male from the waist down," Tracy says. "Once she started wearing girl clothes, she turned into a different person. Her speech changed and she began talking more. If she's posing for a picture, she'll stick a hip out. When she dances, she moves like a girl and flutters her hands. She became a happier person." No longer socially reserved, Barry, now four, enjoys going to preschool, playing with friends, and having tea parties. "We're one hundred percent accepting of her no matter who she is," Tracy says, "but this isn't anything I would wish on my child because of the challenges she's likely to encounter in the world."

In contrast to Barry's early-onset gender dysphoria, clinicians have recently noted a phenomenon in which teens experience a sudden or rapid onset of gender dysphoria (sometimes referred to as ROGD) for the first time during or after puberty. On the upside, the rise of social media has provided teenagers who are grappling with gender identity or gender dysphoria issues with a way to find kindred spirits and support. The downside: some experts are concerned that these online influences may stoke the flames of dysphoria for some people.

In a controversial 2018 online survey, 256 parents who perceived their kids to show signs of a rapid onset of gender dysphoria were recruited from three websites and invited to share their observations by answering ninety questions. Among the kids in this sample, 83 percent were born female, 41 percent came out as nonheterosexual before identifying as a different gender, and 63 percent had reportedly been diagnosed with at least one mental health condition (such as anxiety, depression, or an eating disorder) or a neurodevelopmental disorder (such as attention deficit hyperactivity disorder or autism spectrum disorder) before the recognition of their gender dysphoria, according to their parents.

This study has stirred controversy because parents, not the kids, were asked these questions; also, because an element of social contagion could be at play. Yet another aspect that has caused discomfort is the researcher's conclusion that other factors appear to play a contributing role in gender dysphoria, including a mental health condition, a sexual- or gender-related trauma, a desire to escape one's emotions and difficult realities, a major family stress such as divorce or the death of a parent, or a high level of parent-child conflict.

As Arjee Javellana Restar, a Brown doctoral student and trans advocate, noted in a 2019 critique of this research in *The Archives of Sexual Behavior*, "The majority of methodological and design issues stem from the use of a pathologizing framework and language of pathology to conceive, describe, and theorize the phenomenon as tantamount to both an infectious disease ('cluster outbreaks of gender dysphoria') and a disorder ('eating disorders and anorexia nervosa')." Many transgender activists agree with Restar's perspective, some believing the survey methodology and analysis further stigmatize the experiences of gender-nonconforming youth.

Another wrinkle: some prepubescent children who present as transgender will no longer be gender dysphoric by the time they reach adolescence and will later identify as cisgender. This is called *desistance*, and it's often used as an argument for discouraging social or hormonal transition in these children. It's also a potentially loaded term because in the field of criminology, *desistance* means the cessation of offensive or antisocial behavior. Interestingly, those who go on hormone treatment and transition socially are likely to have a higher *persistence* (or permanence) of their transgender identity, notes Sheri Berenbaum, PhD, a professor of psychology and pediatrics at Penn State University. But it isn't clear whether this is because these actions allow kids to be who they really are or push them to essentially pick a lane by assuming a binary identity.

It took Ben a long time to come to terms with his gender identity.

Born female, he says he always felt different and struggled to fit in—as a child he enjoyed climbing trees, playing volleyball, and playing with building sets. He had dolls, but he was more interested in taking them apart to see how they worked than in playing with them.

At nineteen, Ben married, and at twenty-five he and his husband tried to get pregnant, to no avail. The marriage failed, and after the couple divorced, Ben had a series of relationships with men and three short affairs with women. That's when he began going to therapy and eventually opened what he calls "the gender can of worms." In an effort to feel more empowered, he took up martial arts and boxing. But nothing helped. Because Ben's periods had always been long and painful, as well as emotionally distressful, the therapist suggested taking a break from menstruation. So Ben began taking Depo-Provera, an injection of progesterone every three months, to regulate his periods, but the drug made him feel worse physically. So he began taking a low dose of testosterone to counter the side effects of Depo-Provera. The infusion of testosterone "was like a warm bath—it felt like this was the right chemical in my body," Ben says. Before that, "I felt like I had had estrogen poisoning on the inside."

This physical change, along with all the feelings he'd been grappling with, helped Ben realize that he was transgender. He was thirty-nine when he went on testosterone therapy and eventually had his breasts and uterus removed. These days, he identifies as a gay man and is happily married to Ed, who has long lived as a gay man. Now fifty-six, Ben, a counselor and educator in New York City, says, "I feel lucky that I made it through this journey and that I'm happy and at peace in my life and my body."

The Blurring of Binary Boundaries

Defining gender and sexuality is without doubt a complex challenge, with many nuances and facets, some of which are physical. Some researchers are suggesting that, along with the fish, frogs, and reptiles that are being born with ambiguous genitalia, an increasing number of children are being born with intersex variation, including ambiguous genitalia. Use of the term *hermaphrodite* is perceived as demeaning, which is why *intersex* was introduced as a replacement; more recently, *disorders of sex development* (DSD) has become the preferred medical term.

But reliable statistics on the prevalence of intersex variations are hard to come by, partly because researchers don't always agree on what defines *intersex* in human beings. The term is generally used to describe a variety of conditions in which someone is born with reproductive or sexual anatomy that doesn't conform to the usual definitions of male or female. Seems simple enough, right? Not necessarily, because these anomalies can include abnormalities of the external genitals, the internal reproductive organs, a discrepancy between the external genitals and the internal reproductive organs, sex-chromosome abnormalities, or other unusual conditions.

For example, someone who is born with genitals that seem to be somewhere between the typical male and female anatomy—perhaps an unusually large clitoris or the absence of a vaginal opening on a "girl" or a very small penis or divided scrotum that looks more like labia on a "boy"—could be considered intersex. The same is true of babies who appear female on the outside but have primarily male anatomy on the inside, as well as those whose cells vary between XX chromosomes and XY chromosomes. The category also includes those who are born with congenital adrenal hyperplasia (CAH), an inherited disorder that results in low levels of the stress hormone cortisol and high levels of androgens (male hormones), causing masculinization of the genitals in female infants and early puberty in both boys and girls. Some people aren't discovered to have intersex anatomy until they reach puberty or find out that they're infertile. And "some people live and die with intersex anatomy without anyone (including themselves) ever knowing," according to the Intersex Society of North America.

It's hard enough to define *intersex*, let alone identify the prevalence of these conditions. Based on the instances when doctors at medical centers deliver a baby with noticeably atypical genitalia, the incidence of intersex babies is estimated to be approximately one in fifteen hundred births. But many other babies are born with subtler sex-anatomy variations, which may go undiagnosed. Indeed, experts at Children's National Health System claim that DSDs of some form affect approximately one in a hundred newborns. At this point, determining how common these conditions are is a bit of a guessing game.

Nevertheless, some researchers are wondering if EDCs and other chemicals in the environment could be having an effect on intersexuality of one form or another. After all, research has found an association between high prenatal exposures to EDCs—for instance, if a parent had occupational exposures to pesticides or phthalates—and a higher risk of external genital malformations in male newborns. And researchers at the University of North Texas have explored the physiological pathways through which EDCs can influence sexual differentiation in humans.

Remember, a fetus carrying the Y chromosome becomes a phenotypic male if the testes produce sufficient amounts of androgens at the right time during gestation; if endocrine-disrupting chemicals interfere with this process, the fetus will essentially develop into a female (the default gender, biologically speaking) or develop ambiguous genitalia (that is, have elements of both male and female reproductive organs). As the University of North Texas researchers noted, these chemicals can interfere with the complex biochemical pathways of the brain, which could affect "the way a person associates with his/her physiological sex or personifies his/her gender behaviorally."

From animal studies we have proof of the principle that hormone exposure in utero affects sex-related physical and neural development. Research has shown, for example, that the sexual behavior of rodents depends on the sex of their immediate neighbors in the womb. A female pup who develops between two male pups in utero receives a small extra dose of testosterone from each of her neighbors; as a result, her genitals are somewhat more masculine and when she becomes sexually active, she is more likely to mount other females and less likely to be attracted to males. In another study, male monkeys that were exposed to bisphenol A

(BPA) in the womb were found to exhibit more female behavior, such as clinging to their mothers and social exploration, after birth. In principle, it doesn't matter where the hormone comes from—whether it's from chemicals or natural hormones—in utero; the same changes in genital development and gender-specific behavior can result.

With humans, there are still many unknowns about whether in utero exposure to certain chemicals can affect people's gender identity as they grow up. But this is what we do know: prenatal exposure to endocrine-disrupting chemicals seems to affect the way boys play. In one of my studies we asked moms about how their four- to seven-year-olds played, using a standard "play behavior" questionnaire, and we found that boys who were exposed in the womb to higher levels of the potent chemical di-2-ethylhexyl phthalate (DEHP), which can lower fetal testosterone levels, scored significantly lower on the "masculine scale"—in other words, they were more likely to play with dolls and less likely to play with trucks and guns. Similarly, a 2014 study from the Netherlands used the same play-behavior questionnaire and found that exposure to dioxins and PCBs was associated with more feminine behavior in boys, whereas in girls, exposure to these chemicals was associated with less feminine play behavior.

Meanwhile, research involving females who are born with CAH, which results in their being exposed to high levels of androgens in their early years, has found that even though they are raised as girls, they often exhibit some behavior that is more male-typed. They're not as masculine as "typical" males are, but they are more so than "typical" females. During free-play sessions, girls with CAH, ages two and a half to twelve, chose to play more with boys' toys, particularly trucks, than girls without CAH, and they showed less interest than girls without CAH in classically girls' toys (such as dolls). They're also slightly more likely to have gender dysphoria or to identify as less female, Dr. Berenbaum says. "But the overwhelming majority of girls with CAH identify as girls."

So what does all this mean in the context of this book? Simply, that in addition to influencing the physiology of reproductive development, environmental chemicals may be affecting gender identity and sexual preference. These forms of flux aren't inherently good or bad, but they may present a silver lining: with such trends arguably on the rise, we, as a society, are gradually becoming more open-minded toward accepting people, however they present and identify in terms of gender. That is inarguably a good thing, as we move toward creating a brave, new, inclusive nonbinary world.

Part II

The Sources and Timing of
These Shifts

5

WINDOWS OF VULNERABILITY:
Timing Is Everything

Getting with the Program

Despite being microscopic in size, sperm are mighty and resilient swimmers. These tadpole-like cells are able to recover from numerous forms of environmental assault, dodge and weave their way through various obstacles (hello, cervical mucus!), survive arduous treks through the male and the female reproductive tracts, and exert powerful genetic influences on the developing embryo. Yet, they're also surprisingly vulnerable, particularly during critical periods in a male's development.

While damage to these delicate, hardworking "animalcules" (as Antoni van Leeuwenhoek referred to them upon first viewing them under a microscope in 1677) is possible at any point in a man's life, there are times when a male is especially vulnerable to losing or damaging sperm. These risky periods occur when the germ cells (the primordial cells that will mature into sperm), or the sperm themselves, are rapidly dividing, proliferating, or differentiating. The most sensitive time frame for reproductive tract development is the first trimester of pregnancy, when the genitals and the germ cells that will produce sperm are being formed—a phase called the reproductive programming window. The period between two and four months of age, often called minipuberty because of the early postnatal surge of androgens, including testosterone, is also thought to be highly sensitive to outside influences. Interestingly, testosterone levels peak at the end of minipuberty and then decline to minimal levels by six months. After that, they remain low until shortly before real puberty.

The reproductive programming window is vital for sex differentiation in a developing fetus. A baby's biological sex is determined at conception, based on the specific pair of chromosomes that are present—XX for female, XY for male. Early in the first trimester of pregnancy, the embryo's genital tract looks the same whether the fetus is male or female; it's the same long ridge of tissue. The primordial gonads are just waiting for their operating instructions—the chemical messages that will tell them whether to evolve into male or female genitalia. Approximately eight weeks after conception, these uncommitted gonads begin to undergo big changes, gradually becoming male or female in structure and function, depending on hormone production. Internally, the baby's gonads will become ovaries or testicles. Externally, the fetus either develops a clitoris or the tissue elongates and becomes a penis, and the genital folds become either labia or scrotum. Which way the genitals develop (and how completely) depends on whether testosterone, and how much, is present during this time. In embryos with a Y chromosome, testosterone will be on duty and male-typical sex organs will develop. In the absence of testosterone, female reproductive organs will form.

Looked at another way, female is the default sex for human beings; it's the body's go-to biological sex unless certain hormones swing into action to masculinize the reproductive organs and the brain. To become male, the previously uncommitted genitals need to develop into testicles, the scrotum, the penis, and other male organs; meanwhile, the testicles need to produce enough testosterone at the right time to complete the journey to physical masculinity. The amount of testosterone that's present in a male fetus after the second month of pregnancy is a major factor in determining the size of his penis and other parts of his genitals at birth. By the twenty-second week of pregnancy, the testicles have formed in the abdomen and already contain immature sperm; before long, they'll begin their gradual descent to the scrotum, reaching their ultimate destination late in pregnancy and, in some boys, even after birth.

Any influences that change the production of key hormones during the development of these sexual organs will result in anatomical alterations that are profound and permanent. Such interruptions to the regularly scheduled program can lead to results such as low sperm counts, ambiguous genitalia, shorter anogenital distance (AGD), and genital birth defects such as undescended testicles. For all these parts to develop normally at this stage, a highly orchestrated cascade of events requires precisely the right dynamics at the right times. It's like a ballet: The corps de ballet has to come onstage at the right time to avoid bumping into the principal dancers. If the choreography, or its execution, is off, a principal dancer who leaps high into the air, expecting to be caught by a partner, may get hurt if he isn't there to catch her at the right time. The choreography during the development of an embryo's sexual organs is similarly complex; so many factors are involved that it's a wonder the process works at all.

The Master Switches

When it comes to sexual and reproductive development, hormones are like the great and powerful Oz behind the curtain: unseen but mighty. Hormones are master manipulators, given that they influence virtually every cell in the body, as well as various organs. The entire male reproductive system is dependent on key hormones to stimulate or regulate the activity of its cells and organs. The big ones for male reproduction are follicle-stimulating hormone (FSH), luteinizing hormone (LH), and testosterone. The affected organs include the testicles, the penis, the scrotum, the urethra (the tube that carries urine from the bladder to outside the body and expels sperm during orgasm), and various glands (including the prostate). Anything that interferes with the timing or quantity of these hormones during a critical period of development can disrupt the growth of sex organs and/or their functionality.

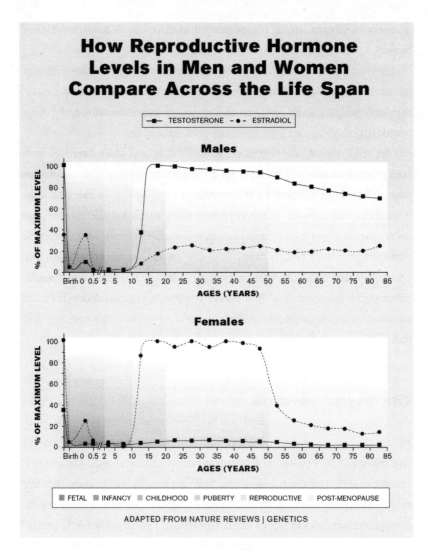

How Reproductive Hormone Levels in Men and Women Compare Across the Life Span

ADAPTED FROM NATURE REVIEWS | GENETICS

The female reproductive system is similarly dependent on hormones, most notably estrogens, progesterone, and testosterone. (Yes, girls and women also produce the male hormone testosterone—it's made in their ovaries, while in males it's made by the testes—but in much smaller amounts than males do.) While they're in the womb, both female and male fetuses are bathed in estrogens produced by the placenta. Once a baby girl is born, her ovaries basically serve as a warehouse for her eggs.

She will also experience a minipuberty that's marked by a hormone surge between two and four months of life, but levels of her sex hormones are much lower than those in boys. As real puberty kicks in, the pituitary gland stimulates the ovaries to start producing estrogen and progesterone, which in turn leads to the onset of menstruation and sexual maturation.

As you've read, a girl is born with all the eggs she's ever going to have—approximately 1 million to 2 million immature eggs, nestled in fluid-filled sacs (follicles) in the ovaries. That may sound like an astonishing amount, and it's certainly more than a woman will ever need, but even that starting point represents a downward trajectory because female embryos may have had as many as 6 million or 7 million eggs while in utero. This is in stark contrast to the male reproductive experience: sperm production occurs in multiple stages from early prenatal development and continues throughout adulthood, with a healthy man producing at least 1 billion sperm per month.

A person's lifestyle habits, as well as certain chemicals that are ubiquitous in the modern world, can hijack the human hormonal system at different times in life. If it happens while an embryo is in the womb, the exposure can create a ticking time bomb that could explode into genital abnormalities, fertility problems, and other health disorders during the person's life. For example, if a woman is exposed to chemicals that block the action of androgens in the first trimester of pregnancy—during what's called the reproductive programming window—it can affect the reproductive development of the male fetus in numerous ways. One is to shorten the anogenital distance (AGD), the span from the anus to the base of the penis, which is significant because research has shown that a shorter AGD correlates with a lower sperm count and a smaller penis. Moreover, prenatal disruption of the male hormonal system can result in reduced testosterone levels and increase the risk that a baby boy will have undescended testicles (cryptorchidism) or a particular type of malformed penis (hypospadias) at birth.

In parallel with declining sperm counts, the incidence of male genital abnormalities has been increasing in some Western countries. Studies from the UK show the incidence of undescended testicles nearly doubled from the 1950s to the early 2000s, while the rate increased more than fourfold from 1959 to 2001 in Denmark. Similarly, from 1990 to 1999, the occurrence of hypospadias, the misplaced opening of the urethra on the penis, increased in Sweden for no discernible reason, and its prevalence more than doubled between 1977 and 2005 in Denmark. As the boys who are born with these abnormalities grow into adulthood, the underlying hormonal havoc can lead to an increased risk of testicular cancer, infertility, and lower sperm count—a bequest most mothers would do anything to avoid giving to their sons.

Samantha, an education specialist, and her husband have been grappling with these worries ever since their son, Ethan, was born in 2018. A twenty-week ultrasound scan, a much-anticipated pregnancy milestone, found Ethan's kidneys to be larger than they should have been. After he was born, he had blood in his diaper when he was four days old from a raging kidney infection that required him to be hospitalized for ten days so he could get intravenous antibiotics. A pediatric urologist told Ethan's parents that his testicles had not descended the way they were supposed to, which puts him at higher risk of fertility issues and testicular cancer down the road—a bombshell of upsetting news for any new parent.

Fortunately, one testicle eventually descended naturally. When he was seven months old, Ethan needed surgery to bring down the other one, which was a centimeter off from where it was supposed to be. Neither side of the family had a history of cryptorchidism, and Samantha says she "led a pristine lifestyle" during the pregnancy, sticking with healthy, organic food and regularly vacuuming with a HEPA filter. So she hasn't been able to figure out why this happened to their son, even after doing extensive research on the subject. "The only thing I can think of is, we live in the Central Valley of California, where the air is bad and we're surrounded by toxins and chemicals," says Samantha, who gave birth to Ethan when she was twenty-four. "It makes me really sad to think that he might not be able to have children if he decides he wants to because of a small problem we got fixed when he was a baby."

While in utero, female embryos' developing reproductive organs aren't quite as vulnerable as those of male embryos are. But this doesn't mean trouble can't happen. Evidence suggests that some of the same chemicals that can affect male genital development in the womb can impact the timing of puberty in girls, leading most notably to earlier development of pubic hair, breasts, and the start of a girl's period. In addition, in utero exposure to some of these same chemical culprits can have a negative impact on a female embryo's ovarian function, leading to a hastened depletion of eggs when she's a grown woman and an earlier age of menopause.

One way or another, what happens in the womb doesn't stay in the womb. These exposures can have long-lasting effects on the reproductive and sexual development of men and women alike.

The Sensitive Male

As far as gender equity goes, the womb doesn't provide a level playing field. This is true regarding potential threats to developing male and female embryos and to the very survival of the fetus. For starters, severe placental dysfunction is more common in pregnancies with a male fetus, which may partly explain the increased risk of early-pregnancy loss of male fetuses.

There's evidence that women's bodies spontaneously abort more male babies during stressful times. For example, the ratio of male to female live births declined in the three to five months after five different terrorist attacks around the world between 2001 and 2012. The extent to which male embryos are chromosomally vulnerable or susceptible to damage from environmental chemicals for other reasons remains to be determined.

Another factor: male fetuses grow faster in the womb, which puts them at greater risk of being undernourished. Insufficient nourishment of the fetus can lead to low birth weight. Also, the risk of preterm birth (being born early) is higher for baby boys. The problem is, male babies that are born at low body weight and/or prematurely are less likely to survive than female babies born at the same weight or week.

No Innocence in the Womb

The womb, of course, is the uterus, and attached to its wall during pregnancy is the placenta. This vital, yet temporary, organ functions a bit like a life-support system for the fetus, providing oxygen, hormones, and nutrients and removing waste products from the fetus's blood. Yet, surprisingly, the placenta isn't as well understood as you might expect.

For example, it was long believed that the placental barrier, a membrane that separates the expectant mother's circulation from the fetus's, was like a wall or a moat that protected the fetus from bacteria, chemicals, and other potential threats. This belief even informed some of the health recommendations that were given to pregnant women— in the 1940s and 1950s, pregnant women were often encouraged to smoke to "calm their nerves" or control weight gain, and champagne and wine were prescribed to treat morning sickness and help expectant mothers relax. These suggestions long ago went the way of the dinosaurs.

Fortunately, our insight into how the placenta works has improved. We now know the placental barrier is far from impermeable, and nicotine, alcohol, and other toxic chemicals such as mercury (from consuming certain fish) can cross or damage it and harm the developing fetus. It's not just that an expectant mother is eating for two: everything she swallows or inhales can potentially affect her baby.

This was discovered the tragic way after diethylstilbestrol (DES), a synthetic form of estrogen, was prescribed to pregnant women between 1947 and 1971 to prevent miscarriage and other pregnancy complications. Later, it was discovered that the adolescent daughters of women who took DES during pregnancy had an increased risk of a rare vaginal and cervical cancer that had never before been seen in young women; they also had higher rates of fertility problems, miscarriage, premature delivery, and ectopic (or tubal) pregnancy, which isn't a viable pregnancy and can be life-threatening to the mother. Long recognized as an endocrine-disrupting chemical, DES has not been prescribed for use during pregnancy since 1971.

Identifying windows of opportunity for detrimental influences on reproductive development is difficult, especially in humans. It's considerably easier in laboratory animals. For example, once it became clear that prenatal exposure to certain environmental chemicals, especially those that can lower testosterone, affects how the genital tract develops, scientists could intentionally vary *when* a pregnant animal was exposed to these chemicals to see how the timing affected the development of male genitals. In rats, researchers found that if an expectant mother is exposed to phthalates—endocrine-disrupting chemicals found in our food, plastics, and other everyday products—eighteen to twenty-one days *after* mating, her male pups' testosterone levels will decrease, and disruptions to normal male genital development will ensue. (When these changes were recognized, they were considered so important that they were given a special name—*phthalate syndrome*.) But this is where things get tricky: if the phthalate exposure occurs only before day eighteen or only after day twenty-one, the syndrome doesn't occur. So these chemicals have a relatively narrow window of opportunity to do their damage in the womb.*

Because it would be ethically unacceptable to conduct a study in which women are intentionally exposed to potentially harmful chemicals during pregnancy, we have had to take a different approach to try

---

* Keep in mind that the gestational period and the reproductive-development time frame in rats are quite different from those in humans. A rat pup is born after spending about twenty days in the womb, whereas a human baby spends an average of two hundred and eighty days in utero. After birth, rats' bodies, including their genitals, are considerably less developed than human babies' are. Another major developmental difference: a rat enters puberty about forty days after birth, whereas humans typically take eleven to twelve years to reach that formative milestone.

to ID the sensitive window for phthalate exposure during pregnancy in humans. In studies my colleagues and I conducted between 1999 and 2009, we examined the effects of a pregnant woman's *incidental* exposure to phthalates on the reproductive development of her male offspring. We did this by measuring levels of these chemicals in the expectant mother's urine during various stages of her pregnancy. When we looked for the phthalate syndrome and the programming window for the development of male genitalia, we found that it occurred in the second half of the first trimester, specifically during weeks eight to twelve of pregnancy. When we examined these boys after birth, we found that the anogenital distance was shorter and the penis was smaller than expected for a boy of his size whose mother had lower exposure to certain phthalates.

Remember, at the same time that testosterone is causing the penis to form in a male embryo, the predominantly male hormone is increasing the length of the male AGD. If not enough testosterone is present during this key time period, a baby boy might end up being born with a shorter AGD, a smaller penis, and testicles that are less completely descended, as my research team first revealed in 2005. Guys are right—size matters when it comes to their genitals, just not in the way they think it does. In terms of fertility, the length of the AGD is more significant because a shorter AGD is linked with a shorter penis size and a lower sperm count. After my research was published, I was deluged with emails from men asking if their AGD was long enough, and from women who were worried about whether their use of phthalate-containing cosmetics during pregnancy could have affected their sons' AGD or sexual development. I tried to be helpful, but it's difficult to draw a causal connection between reproductive development and a particular culprit in any one instance, especially in retrospect. In this case, hindsight isn't 20/20.

The AGD is such an important marker of reproductive health and endocrine disruption that it should probably be measured in every infant. But it still isn't among humans, outside of the research realm. I think of AGD as a bit like Janus, the ancient Roman god of beginnings and transitions who is depicted with two faces, one looking to the future and one to the past. The length of a baby's AGD can tell us what chemical influences the fetus was exposed to in the womb, as well as what the future holds for that person's reproductive health and fertility; thus, the AGD offers a rearview-mirror perspective and a forecast of the person's future health.

Yet it continues to amaze me that no one pays attention to the AGD. Admittedly, it's an awkward subject to talk about in polite company. Few adults are familiar with the phrase or the acronym, though kids use various slang terms, such as *gooch* or *taint*, to refer to anogenital distance; but they have little appreciation of how significant its length is.

By any name, the AGD is the body part that differs most in size between the sexes. It's usually 50 to 100 percent longer in males than females, even after adjusting for relative body size. In women, the AGD represents the distance from the center of the anus to the top of the clitoris—and it means something for girls, too. If a female embryo is exposed to too much testosterone in utero—which can happen if the mother has polycystic ovary syndrome (PCOS)—she will be born with a longer-than-usual AGD for her sex. Put another way: AGD can be viewed as a biological marker of prenatal androgen activity, and given the association between a longer AGD and PCOS in girls, it appears that PCOS may originate in the womb.

Exposure to certain environmental chemicals can also have androgenic effects in the body, although these are rare compared to the number of chemicals that lower androgen. Research has found that liquid waste products from pulp and paper mills demonstrate androgenic activity, "often of sufficient potency to masculinize and/or sex-reverse female fish," according to the Environmental Protection Agency. In a science fiction–like feat, many species of fish have the capacity to change their gonads and secondary sex characteristics, such as pigment or body shape, during adulthood. This doesn't happen naturally or randomly but in response to environmental stimuli such as changes in water temperature affecting wildlife or the presence of pharmaceuticals that can alter hormone levels (more on that in chapter 9).

The Long and the Short of Exposures

As you've seen, changes to the developing fetus's reproductive system can literally last a lifetime. For example, a decrease in the number of male germ cells that occurs as a result of mom's—or dad's—smoking can affect their son's semen quality as an adult. By contrast, if the chemical exposure occurs later in life, the changes are reversible. A grown man who smokes cigarettes typically experiences a 15 percent decline in his sperm count, an effect that can be reversed if he quits the habit; however, if an expectant mother smokes during pregnancy, her grown son may experience a fairly dramatic decrease in his sperm count—up to 40 percent—that is *irreversible*. It's not only chemicals that can have negative effects. New research suggests that if an expectant mother experiences significant life stress—such as job loss, divorce, or the death or illness of a loved one—early in a pregnancy with a male fetus, her son is at increased risk for having reduced sperm count, fewer progressively motile sperm, and lower testosterone levels at age twenty.

Scientists use the terms *organizational effects* and *activational effects* to distinguish between these types of influences. Organizational effects occur early in an individual's lifetime and induce permanent alterations to the structure and function of cells, tissues, and organs. By contrast, activational effects are usually rapidly occurring but transitory influences that happen during adulthood. Sounds simple enough, right? Well, complicating matters, some of the same sex hormones and endocrine-disrupting chemicals can have either organizational or activational effects on embryos, fetuses, children, or adults, depending on when the exposure occurs.

Intuitively, it might seem as though only high doses of chemicals are likely to be problematic. But the reality is, embryos are sensitive to low doses of environmental chemicals because they are small and undergoing high rates of cell division. We're talking about amounts that may be as small as a drop of baby oil in an Olympic-size swimming pool—nearly minuscule. Nevertheless, if a pregnant woman—and hence her developing baby—is exposed to low doses of certain chemicals at sensitive periods in the embryo's reproductive and neural (brain) organization, the effects can be substantial and permanent. That's right—it's not only the reproductive organs that are affected. During periods of gestational development, when sex hormones exert organizational effects on the fetus's brain, a mother-to-be's exposure to endocrine-disrupting chemicals can affect her offspring's patterns of behavior that are considered traditionally male or female later in life.

An interesting example in animals: In experiments with rats, researchers exposed male and female rats to a class of endocrine-disrupting chemicals called PCBs both while they were in their mother's wombs and again when the rats were juveniles. The doses of PCBs were comparable to what humans experience in the real world, and the rats' development was followed as they matured in life. The researchers found that both prenatal and juvenile exposures to PCBs had significant effects on the rats' expressions of anxiety or aggression, as well as on their sexual or risk-taking behaviors. Interestingly, the juvenile exposures magnified the effects of prenatal exposures on anxiety-related behaviors—in other words, when the rats were exposed both times, the changes were more pronounced; an additive effect occurred.

These effects are in line with what's called the two-hit model of disease development. In simple terms, when it comes to cancer, the model suggests that two "hits" to DNA are necessary to cause the disease. The first hit can stem from a genetic mutation, while a subsequent hit could come from environmental exposures and other nonhereditary factors. In terms of reproductive tract and brain development, it's now recognized that the first hit can happen in the womb, and a second or third hit can happen during a baby's early months, during puberty, or even during adulthood. The two-hit model is the developmental equivalent of adding insult to injury. Over time, toxic influences can have cumulative effects on reproductive development and function, leading to potential fertility problems or other health challenges long before a man or a woman is even contemplating having children.

It's no secret that at puberty kids often engage in risk-taking behaviors.

The substances and chemicals they're exposed to could have lasting repercussions for their health because these can affect the development of a teen's brain and reproductive system. This is at least partly because puberty is a period of continued neural sensitivity to the organizing effects of hormones. During adolescence, for example, teens are particularly sensitive to the effects of alcohol and smoking, and research has revealed that early alcohol consumption (as early as sixth grade) can delay pubertal development. Developing breast tissue in girls is susceptible to the effects of certain phthalates, leading to increased breast density; pubertal gynecomastia, breast formation in boys, has also been linked with higher blood levels of certain phthalates. As far as below-the-belt effects go, sperm are being produced during puberty and are susceptible to the adverse effects of many factors, including chemicals that can alter the young man's hormones or the complicated physiological processes that work together to produce the sperm.

Now that you have a high-altitude view of the precarious periods in the life of a fetus, developmentally speaking, here's the surprising part: These windows of vulnerability aren't new; they've always been there. It's just that we didn't know until relatively recently the extent to which children's sexual and reproductive development could be affected in the womb by their parents' lifestyle practices and chemical exposures, or by their own exposures in early childhood and during adolescence.

Just as timing means everything for conception, timing is paramount in a child's reproductive development. By way of example, consider this: A research group examined the number of eggs retrieved from women undergoing IVF and compared them to the amount of a nonphthalate plasticizer called DINCH in the women's urine. The researchers retrieved fewer eggs from women with higher levels of this chemical. What's interesting is the drop in the number of eggs that were recovered was stronger among women who were over age thirty-seven, compared to younger women, which suggests that as a woman and her partner age, their bodies may become less resilient to the effects of harmful chemicals. Another challenge to add to the list for older parents!

So, with regard to reproductive development, it's not just about *what* you consume; it's about *when* you consume it. If you're a man who smokes before conception, that's a risky proposition; if you're a pregnant woman, the first trimester, in particular, is a delicate time for the fetus's genital development—and the fallout isn't limited to the possibility that your son will end up with fewer sperm or your daughter with higher androgen levels. The potential ripple effects for your future sons' and daughters' sexual and reproductive futures are substantial, as you'll see in a later chapter.

6

UP CLOSE AND PERSONAL:
Lifestyle Habits That Can Sabotage Fertility

The Challenge of Measuring Up

When a man visits a sperm bank to make a donation, certain lifestyle practices can quickly land him on the no-fly list. Use of illicit drugs is an obvious one. The same is true if the aspiring donor takes nearly any medication on a daily basis or has been exposed to or is infected with sexually transmitted diseases (STDs). Many sperm banks also ask about recent illnesses with fever because fever is associated with declines in sperm quality, but these are temporary influences, rather than permanent deal breakers. Certain lifestyle factors can also have such a negative effect on a man's sperm quality that he ends up not making the cut. These include exposure to certain occupational or environmental hazards, smoking, excessive alcohol use, nutrient deficiencies, overheating, and general couch potato habits.

These issues don't even consider the basic requirements for eligibility, which vary slightly from one sperm bank to another. At the California Cryobank, for example, aspiring donors must be at least five feet eight inches,* between nineteen and thirty-eight years old, a college graduate

---

*   Generally, people want their children to have a height advantage, whether it's so they can excel at sports, have an easier time managing their weight, be more (or in college),* in good health, legally allowed to work in the United States, and have sexual partners who are exclusively female. The Sperm Bank of California has similar criteria but is slightly more flexible about height (five feet seven inches is the minimum). The Northwest Cryobank in the Pacific Northwest has the additional requirement that applicants be within normal weight limits for their muscular build and height.

Ultimately, a guy has a better chance of being admitted to Harvard, Princeton, or Yale than he does of being accepted as a donor at the country's leading sperm banks. Some have an acceptance rate as low as 1 percent.

Aside from the aesthetic and educational requirements, which largely stem from client preferences, there are good reasons for many of these highly selective standards: these elements could affect a man's sperm quality and the health of a baby that's conceived. Most sperm banks, for example, won't accept donations from men over forty because an older man is likely to have more DNA damage in his sperm than a man in his twenties or thirties. Certain lifestyle practices can also damage DNA in sperm, as well as compromise sperm concentrations, motility, and morphology. Yet most men are unaware of this.

Fertility-Foiling Lifestyle Factors

The reality is, while we're all going about our daily business, men *and* women could unwittingly be harming their reproductive health and fertility, ignorant of this possibility until they have trouble conceiving. Aspects of the modern diet and lifestyle are bad for sperm, and women's reproductive function isn't immune to these influences. Some lifestyle practices—such

---

sexually appealing, or possibly earn a higher salary. Some studies have found that taller people make more money and have more opportunities for management positions.

*   Among sperm seekers, signs of intelligence are highly prized, especially since sperm can't take IQ tests.

as smoking and heavy alcohol use—won't come as surprises because they're known to be harmful to your heart, lungs, bones, and other areas. But your doctor may not have mentioned—and your mother didn't know—that what's bad for these organs and tissues can be bad for reproductive function, too, kicking up the risks of problems with sperm quality in men, as well as with menstrual function, miscarriage, ovarian reserve, and other reproductive parameters in women.

It is worth nothing that body burdens are slightly different for men and women (spoiler alert: a greater number of lifestyle factors can harm a man's sperm than a woman's eggs) and so is the time frame for these influences to potentially do their greatest damage. Women's reproductive life span lasts twenty-five to thirty-five years, whereas for men it can be much longer (the oldest reported father was ninety-six!). Because sperm are continuously produced during adulthood, men whose lifestyle habits have compromised their semen quality may be able to improve it by changing their behaviors; they get a do-over, a chance to hit the reset button.

Women aren't always as lucky in this respect. It's true that if a woman has exercise-induced amenorrhea (the absence of menstruation) or is underweight because she's not eating enough, exercising less and eating more may restore her estrogen levels to the normal range and her periods back to a more regular cycle, including more consistent ovulation. But with that exception, she has fewer opportunities to potentially reverse the misfortune of reproductive problems that have befallen her.

Here's a closer look at how specific lifestyle-related factors can harm reproductive health.

Body Weight

One factor that has an equal opportunity influence on reproductive function in men and women is body weight. Of course, weight isn't a lifestyle factor, but diet and exercise patterns are, and they can have substantial effects on how much someone weighs. This has little to do with the plastics and chemicals in our midst, although EDCs, some of which have been called obesogens, can influence how much weight we gain. It has a lot to do with the quality of our food choices and our levels of physical activity. There's no denying that it's highly challenging to manage your weight in the modern world, given that high-calorie, processed, and superprocessed foods are within reach nearly everywhere you go. And it's easy to get through a day with little movement now that we're living in the age of automated everything. These realities may be taking a toll on human reproductive function, as well as body weight.

Being substantially overweight or underweight has a negative effect on sperm quality, and obesity (a body mass index, or BMI, of 30 or higher) is especially detrimental because it's associated with lower sperm count, concentration, and volume, decreased sperm motility, and a higher incidence of abnormally shaped sperm. For women, there's also a U-shaped curve when it comes to the link between body weight and miscarriage—women with a BMI of 30 or higher or a BMI less than 18.5 have an increased risk of miscarriage.* Similarly, if a woman's weight is too high or too low, it can affect her chances of getting pregnant because she may not be ovulating regularly or may not have the proper amounts of estrogen and progesterone to support a healthy pregnancy. This is another example of the Goldilocks principle: men and women alike have a sweet spot—or a "just right" zone, in the words of Goldilocks—for body weight as far as optimal reproductive function and fertility go.

Considering these associations, it may not be a coincidence that the decline in sperm counts, the uptick in fertility problems, and the rise in obesity rates in Western countries have occurred in tandem. From 1999 to 2016 alone, the obesity rate among adults in the United States increased by 30 percent, with nearly 40 percent of adults tipping their scales into the obese category in 2016.

---

*   Reality check: when it comes to miscarriage, obesity is much riskier than being underweight.

Smoke Gets into Your Private Parts

As you've heard countless times—including just a few pages ago!—smoking is among the most harmful health habits on the planet. It's also among the most damaging influences on men's reproductive function. Cigarette smoking is associated with reduced sperm count and motility and an increase in defects in shape, with more dramatic detrimental effects in moderate to heavy smokers than in lighter smokers. But any amount of smoking, even exposure to secondhand smoke, is harmful to sperm.

Research in mice found that those subjected to environmental cigarette smoke had sperm with missing tails; this makes it difficult, if not impossible, for the little swimmers to reach the egg. In humans, the chemicals in cigarettes have been found to cause damage to the DNA in sperm, reduce testosterone levels, and impair the sperm's ability to fertilize an egg. (BTW, smoking also increases the risk of erectile dysfunction.)

For women, too, smoking is the most injurious lifestyle factor when it comes to reproductive health. The chemicals in cigarettes—nicotine, cyanide, and carbon monoxide—are toxic to a woman's eggs and speed up the rate at which they die off. Infertility rates are significantly higher among women who smoke, and the risk rises with the number of cigarettes a woman smokes. Smoking also increases a woman's risk of having a tubal (or ectopic) pregnancy or miscarriage—and the amount of time it takes the woman to get pregnant, whether she's trying to conceive the old-fashioned way or through IVF. Moreover, because smoking damages the genetic material in eggs *and* sperm, women who smoke are more likely to have a chromosomally abnormal fetus, such as one with Down syndrome.

Exposure to secondhand smoke is also harmful for women's reproductive function. Research has found that women who are exposed to secondhand smoke often take longer to get pregnant. In addition, women who had never smoked but had the highest exposure to secondhand smoke, whether it was at home as a child or as an adult or at work, had significantly higher risks of having a miscarriage, stillbirth, or ectopic pregnancy. That same group also has an increased chance of going through natural menopause before age fifty. And there's no question that passive smoking (aka, exposure to secondhand smoke) is nearly as damaging to a developing fetus's health as if the mother actually smoked.

While rates of cigarette smoking among adult men and women in the United States have declined by more than 50 percent since 1964, nearly 38 million of them (fourteen out of every one hundred) still light up daily or frequently. Worldwide, rates of cigarette smoking are considerably higher—nearly 20 percent of the world's population smoked in 2014. Smoking rates are slightly lower among women (12 percent) than among men (16 percent) in the United States. But worldwide, men smoke nearly five times more than women, with the highest rates for men found in Western Pacific countries.

Marijuana is the most widely used recreational drug in the United States, and its use continues to grow, especially as more states legalize it. Many younger people, in particular, currently believe that it's safer to smoke weed than nicotine, but it may be a mistake to think that marijuana is less toxic to sperm. There hasn't been much research on this issue, but it's starting to trickle in. A 2015 study from Denmark found that regularly smoking marijuana more than once a week was associated with a 29 percent lower sperm count; even worse, men ages eighteen to twenty-eight who used marijuana more than once a week as well as other recreational drugs reduced their total sperm count by 55 percent. Among men undergoing fertility evaluation as a precursor to assisted reproduction, those who used large quantities of marijuana were four times more likely to have poor swimmers, and moderate users were nearly three and a half times more likely to have abnormally shaped sperm. Women aren't impervious to such harmful reproductive effects. A 2019 study found that women who smoked marijuana when they underwent infertility treatment with ART had more than double the miscarriage rate of those who didn't.

There's also preliminary evidence that using e-cigarettes, or vaping, may damage sperm. Some animal studies suggest that even cannabidiol (CBD), the second most prevalent active ingredient in marijuana, could damage sperm development and reduce the ability of sperm to fertilize an egg, although not much research has been done on this substance. This isn't that surprising since CBD products have only recently become supertrendy. The use of e-cigarettes has also become popular, especially among young adults, with 28 percent of high school students in the United States fessing up to using these tobacco products regularly, according to a 2019 survey of more than ten thousand high school students. How these new trends will affect the fertility of this generation of young adults remains to be determined. Stay tuned!

A Toast to Good Semen

While any amount of smoking is bad news for sperm, semen is more forgiving when it comes to alcohol. Like body weight, this is another variable with a sweet spot: moderate alcohol intake—defined as four to seven units per week (for the record, one glass of wine and one bottle of beer each constitute one unit)—is associated with higher semen volume and total sperm count; but high intakes—more than twenty-five units per week—are hazardous to sperm and other aspects of semen quality. Chronic or excessive alcohol intake may reduce testosterone production, which could compromise sperm production and other aspects of semen quality. And though it's not a consistent effect, some scientific, as well as anecdotal, evidence links heavy alcohol consumption with a greater risk for erectile dysfunction. Guys often refer to this effect as whiskey dick, which *Men's Health* magazine calls "the greatest curse known to mankind."

The same guidelines for alcohol apply to women: stick with moderation. Low to moderate alcohol consumption (one drink per day) before pregnancy does not affect a woman's risk of having a miscarriage or stillbirth. By contrast, binge drinking (for women, tossing down four or more drinks on one occasion) is known to be harmful to the heart, mind, and other parts of the body. Research suggests that frequent binge drinking in women can have an adverse effect on ovarian reserve, given that it's associated with lower levels of anti-Mullerian hormone, which is produced by the ovaries—26 percent lower, according to one study. This is particularly worrisome since the rate of high-risk drinking among women in the United States has been on the upswing, increasing by 58 percent from 2001 to 2013. It goes without saying, of course, that drinking during pregnancy is a major no-no.

Foods for (In)Fertility

A man's eating habits can affect his fertility, for better or worse, too. Some of the most compelling findings about the influence of diet and nutrition on semen quality come from the Rochester Young Men's Study (RYMS), which I've been leading since 2007, and the analyses are ongoing. For RYMS, we recruited male college students who were enrolled at the University of Rochester in New York between 2009 and 2010 and had each man provide a semen sample and complete detailed questionnaires about his own food intake and his mother's eating habits while she was pregnant with him. RYMS was part of a multicenter international study that aimed to evaluate the effects of environmental contaminants on semen quality—and the findings were nothing short of illuminating.

On the negative side of the ledger, a high intake of full-fat dairy foods, especially cheese, was found to be associated with greater abnormalities in sperm quality. These unfortunate effects might be due to the large amounts of estrogens in dairy products or to the presence of environmental contaminants such as pesticides and chlorinated pollutants in these products.

Many people don't realize that hormones, including estrogen, progesterone, and testosterone, are given to beef cattle and sheep sixty to ninety days before slaughter to promote their growth, and residues of these hormones persist in the meat. One of our studies found that when pregnant women ate seven or more beef-containing meals per week, their sons had reduced sperm counts. Meat processing—such as salting, curing, fermentation, and smoking—is also of concern. Men who eat a lot of processed meats (think hot dogs, bacon, sausage, salami, and bologna) tend to have a lower sperm count and a lower percentage of normally shaped sperm. In addition, the curing of meats produces chemicals, including nitrates and nitrites, that can cause cancer and also damage DNA, including DNA in sperm.

Healthy young men who are lean but drink more sugar-sweetened beverages, such as sodas, sports drinks, and sweetened iced teas, have reduced sperm motility, compared to men who rarely consume these drinks. That these effects were confined to lean men, rather than overweight or obese men, suggests that they may be due to the promotion of insulin resistance and oxidative stress, which are known to negatively influence sperm motility.

Long before a woman is eating for two, her diet may affect her reproductive health and functionality. For a woman's fertility, a high intake of meat and trans fats is among the biggest dietary demons. On the positive side, an adequate intake of folic acid is not only important during pregnancy (since it can prevent neural-tube defects such as spina bifida in the baby), but a higher intake before conception may also increase a woman's chances of becoming pregnant and decrease her risk of miscarriage.

Women who can't imagine giving up their morning cup of java can rest assured: this habit isn't damaging to female fertility, ovarian function, or other aspects of reproductive health. But *moderation* is the watchword here because there are hazards associated with overdoing it. For one thing, consuming too much caffeine during pregnancy can be problematic—a couple of cups of coffee per day aren't, but downing four or more servings per day is associated with a 20 percent increased risk of miscarriage and giving birth to smaller-than-expected babies.

Couch Potato Habits

Spending long hours binge-watching favorite TV shows may be a feel-good way to unwind, but it won't do a man's semen any favors. In a study involving 1,210 healthy young Danish men, researchers found that long periods of television watching were associated with dramatically lower sperm counts and decreased testosterone levels. The sperm concentrations of men who watched TV more than five hours per day were 30 percent lower than those of men who didn't tune into the tube at all—but a decline was seen for any amount of TV watching.* These effects may be due in part to the increase in the scrotum's temperature that comes from sitting still; increased scrotal temperature temporarily reduces sperm production. Interestingly, the same effects were *not* found for men who worked long hours at a time sitting at a computer. So the full story is still a bit of a mystery.

Another Move-It-or-Lose-It Effect

Among adults in the United States, physical activity trends have been heading along a healthy trajectory, with a 24 percent increase from 2008 to 2017 in the number of adults meeting the guidelines for minimum aerobic exercise (150 minutes of moderate-intensity or 75 minutes of vigorous-intensity exercise per week). Those are certainly steps in the right direction (pun intended), but there's still plenty of room for improvement because 46 percent of adults are *not* getting the recommended dose of movement. Regular physical activity is beneficial for reproductive function, as well as cardiovascular and brain health.

An exception to this move-it-to-boost-it dynamic: bicycling. Men

---

* Remember when "Netflix and chill" became code for casual sex, as it referred to watching TV with a sexual prospect? These days, a new connotation may be in order: it may mean simply relaxing and watching a movie—while putting your sex life on ice.

who reported cycling for ninety minutes or more per week had 34 percent lower sperm concentrations than those who didn't ride bicycles at all. Another study examined the influence of cycling on sperm qualities and found that long-distance competitive cyclists had less than half as many normally shaped sperm as their less active peers did.* One theory here is that a hot and bothered scrotum can cause deleterious effects on sperm production, while another suggests that compression from the seat against a man's private parts can affect blood flow to the testicles.†

Among the biggest potential lifestyle-related threats to a woman's reproductive health is the triple whammy of eating too little, exercising too much, and having menstrual irregularities. This is a big deal for several reasons, chief among them: if a woman doesn't have periods (meaning she has amenorrhea) or has highly irregular menstrual cycles, the level of estrogen in her body may be lowered significantly. Naturally, this is a problem if she wants to have a healthy pregnancy. But the low estrogen also causes her to lose bone density and strength, which can put her at risk for stress fractures and osteoporosis.

The combination of disordered eating (including full-blown eating disorders, subclinical ones, and excessive exercise), menstrual dysfunction, and low bone density can lead to what's called the female athlete triad. While any physically active woman can develop one or more parts of the triad at any age, those at greatest risk include women who participate in physical activities that place a premium on appearance or that prize endurance. In the aesthetic category are cheerleading, dance, figure skating, and gymnastics; in the latter are sports such as distance running or rowing.

---

* Before hanging up the bike for good, however, guys who want to achieve a pregnancy should know that some fertility experts believe that modifying the height and shape of the bike seat, and the geometry between the seat and the handlebar height, can reduce the pressure that's placed on the genital area and improve sperm parameters.

† Some other ways the heat can catch up with guys' crotches: regular use of saunas and hot tubs is correlated with a drop in sperm counts and motility. Fortunately, all of these effects appear to be reversible once men stop engaging in these hot recreational activities.

Even without the other elements of the triad, extreme exercise—as in exercising to the point of exhaustion—on a daily basis more than doubles a woman's risk of having ovulatory dysfunction and infertility. This is at least partly because excessive amounts of exercise can lower hormone levels and cause a woman not to ovulate or to ovulate irregularly. By contrast, moderate exercise, defined as physical activity that's performed at a moderate intensity for less than an hour per day, is associated with a reduced risk of infertility. In other words, moderate exercise is a healthy source of physical stress, whereas excessive exercise tips the balance into overload territory.

During graduate school, Susannah took her occasional jogs to the next level, cranking up their frequency, pace, and distance. She had lost fifteen pounds the previous summer and was deluged with compliments about her newly slender five-foot-nine-inch figure. Because she was worried about gaining back the weight despite running twenty-five to thirty-five miles per week, she began skipping meals or eating very lightly and sometimes even purged or doubled up on her runs after eating too much. The result: Susannah lost seven more pounds—and her period. "I was secretly thrilled to not have the hassle of my period, but after five months, it came back with a vengeance, every two to three weeks, and that was a nightmare," she recalls.

That's when Susannah saw her doctor, who diagnosed her with an exercise-induced hormone disorder and warned that she was putting herself at risk for bone loss and a stress fracture; the doctor didn't mention fertility problems as a possible consequence, but Susannah later found out they could have resulted. The doctor advised Susannah to either cut back on running and gain some weight or to take oral contraceptives to regulate her menstrual cycle. By then, she was addicted to running, so she chose the latter option—until she discovered that the Pill gave her headaches and extreme breast tenderness.

"It was a tough trade-off because I loved being thinner, but I couldn't stand the way the hormones made me feel," she recalls. So she stopped taking the oral contraceptives and began limiting her running to four times per week, and eating regular meals again. Within three months, she'd gained eight pounds and her periods resumed a regular pattern.

Stress and Fertility

It may be anxiety provoking to recognize the extent to which lifestyle factors can affect sperm production and fertility, but we haven't even gotten to the issue of stress. Besides affecting a man's state of mind, the unavoidable stresses and strains of modern life can take a toll on his sperm production. This is especially true if his personal stress meter registers *overload*, which can happen quite easily these days.

In a study of 1,215 Danish men, researchers found that those who reported the highest stress levels on a psychosocial questionnaire had 38 percent lower sperm concentrations than men who reported intermediate stress levels. Some of my own research has found that men who've experienced two or more recent stressful life events—such as the death or serious illness of a close relative, divorce or serious relationship problems, moving, or a job change—were more likely to have below-normal sperm concentration, motility, and morphology. And medium and high levels of work stress have been associated with sperm DNA damage. One way or another, experiencing excessive psychological stress can essentially put an OUT OF ORDER sign on the sperm production machinery, not to mention a man's sex drive.

The complicated issue of stress is even worse for women, who are nearly twice as likely to suffer from severe stress as men are. Among other health effects, stress can send a woman's libido packing, just as it can for a man—another rising hazard in the contemporary world that can affect people's reproductive potential. And some research has found that women with high levels of perceived stress are more likely to have irregular or painful periods, and more premenstrual symptoms, which can kill the mood.

All that said, the relationship between stress and fertility isn't quite so simple. For decades, the connection has been hotly debated, and the jury is still out on this. The reason: women who are undergoing fertility treatments, including IVF, report high levels of stress, but it's not clear whether stress itself can cause or contribute to infertility. It's a chicken-and-egg kind of mystery.

Meanwhile, some compelling evidence links high levels of psychological stress to an increased risk of miscarriage, particularly recurrent miscarriage, although this association isn't clear-cut, either. In fact, when researchers from the Naval Health Research Center in San Diego examined whether the military experiences of US servicewomen who were deployed in Iraq and Afghanistan increased their chances of having a miscarriage or impaired fertility upon their return, they found that military deployment (an intensely stressful experience if ever there was one) didn't increase the risk for miscarriage or fertility problems. This is encouraging news for civilian women who are stressed-out and want to get pregnant.

Sex, Drugs, and Reproductive Function

A number of medications, too, can KO reproductive function, particularly hormonal agents and antineoplastic agents, which are used to treat cancer. Others can as well. What isn't widely known about the US opioid epidemic is that these powerful pain-relieving drugs can increase DNA damage in sperm, and with high doses of opioids, testosterone levels drop significantly. Farther down the pain-medication potency scale, Tylenol (the generic name is acetaminophen; it's known in Europe as paracetamol) has been shown to cause sperm abnormalities, including DNA fragmentation, and to increase the time it takes to achieve a pregnancy; moreover, taking high doses of Tylenol can alter the shape of sperm in ways that can compromise their fertilizing capabilities.

Some male athletes use anabolic androgenic steroids, which are synthetic or man-made variations of testosterone, to improve their performance and/or increase their muscle mass and strength. Besides having serious and potentially irreversible adverse effects on various organs and body systems, including the reproductive system, these steroids can throw hormone levels significantly out of whack. If they're overused, these steroids can lead to structural and functional changes in sperm, a reduction in the volume of the testicles, enlarged breasts, and subfertility in men.

Testosterone supplementation is the gold standard for treating patients with male hypogonadism, a condition in which the testicles don't produce enough testosterone. While testosterone replacement therapy helps restore muscle strength, prevent bone loss, and increase energy and sex drive in men with hypogonadism, it often impairs sperm production and can even completely eliminate it in some men. Given the increasing incidence of hypogonadism and the rise in older men who want to have children but don't have enough testosterone to do the job—39 percent of men ages forty-five and older have hypogonadism, according to one US study— health-care providers are increasingly encountering men with testicular failure who want to restore their fertility. That's not a simple proposition.

At every age, women are twice as likely to take antidepressants as men, and the use of these medications increased 64 percent from 1999 to 2014 for both genders. And—are you detecting a pattern?—the use of SSRIs (selective serotonin reuptake inhibitors), which are prescribed primarily for depression or anxiety, reduces sperm concentration and motility and increases the percentage of abnormal sperm.

For women trying to conceive, some evidence suggests that taking antidepressants may reduce the probability of success in a given menstrual cycle by 25 percent. What's more, concern is mounting about drug-induced amenorrhea—menstrual irregularities that are brought on by the use of antidepressants, as well as anti-psychotic and anti-seizure drugs. These effects are complex but worth mentioning, given that the use of antidepressants alone has skyrocketed in the United States. They are undeniably a potent factor that could affect the reproductive health and functionality of millions of women of childbearing age.

Undoing the Damage

The good news is that many of the detrimental effects that I've been telling you about are reversible. After giving up cigarettes, heavy drinking, bicycling, or SSRIs, a man's sperm integrity may improve considerably. Case in point: A few years ago, a twentysomething man who was a regular sperm donor at the Fairfax Cryobank in Philadelphia was put on a break after he experienced a drop in his sperm count and motility and an increase in round cells* in his semen sample. As the staff talked to him about these changes, the donor mentioned that he had moved in with a woman who was a smoker, started a new job that was stressful, and was eating a lot of fast food and junk food. The staff made recommendations for improving his diet, getting more sleep, managing stress better, and minimizing his exposure to cigarette smoke—and sent him on his way. Three months later, he returned and his sperm quality had rebounded to where it was before.

As you've seen, assuming they have healthy sperm to begin with, men are in the enviable position of having the chance to reestablish a clean slate, given that sperm are continually being produced in a process that takes sixty to seventy days. So if men improve their lifestyle habits, they can reset their sperm production. A woman's eggs don't have the opportunity to regenerate the way sperm do; instead, once they're fried, that's it—they're cooked and the damage is irreversible.

All of this is to say, the highly hectic, pressure-packed lives many people lead appear to be taking a toll on their sex drives and fertility. It's hard to determine whether the declines stem primarily from altered hormone levels, increased stress levels, poor lifestyle choices, or other factors. But, one way or another, it's clear that modern life is having a chilling effect on people's reproductive health and well-being.

---

*   Round cells aren't well understood but are currently thought to be immature sperm; they can result from a "spermatogenic insult," even the flu.

7

SILENT, UBIQUITOUS THREATS:
The Dangers of Plastics and
Modern Chemicals

The Promise of Plastics

Remember the cocktail-party scene in *The Graduate* in which Benjamin Braddock, the recent college grad played by Dustin Hoffman, is making the rounds and chatting with guests? At one point, Mr. McGuire, a friend of Ben's parents', takes him aside and says he has one word for him: "*Plastics!* . . . There's a great future in plastics."

After World War II, chemical companies launched campaigns suggesting that plastics could be molded to meet myriad needs and provide greater convenience in modern life. Before long, plastics, and the chemicals they contain, became ubiquitous in water bottles and food packaging, in cars, computers, and other electronic devices, and in other everyday products. In particular, chemicals in plastic include phthalates, which make plastic soft and flexible; bisphenol A (BPA), which makes products hard; and polyvinyl chloride (PVC), which is versatile and can be used in a range of products, including children's toys, building materials, and food packaging. The combination of scant regulation and high consumer demand led to the era of "better living through chemistry."

Plastic remains everywhere in our world—and we're starting to pay a price for its ubiquity. The same is true of pesticides, flame retardants, and other chemicals in widespread use. This despite the fact that Rachel Carson's groundbreaking 1962 book, *Silent Spring*, drew global attention to a mounting concern among scientists and activists that synthetic chemicals were having negative effects on wildlife and the environment and posing health risks to humans. Since then, things have only gotten worse.

One problem is the almost complete lack of regulation of these chemicals. Unlike drugs, which must have a proven record of safety and efficacy before they're allowed to come to market, chemicals are largely presumed innocent from the start—they're considered safe until proven otherwise. This means manufacturers can use these chemicals in a wide array of consumer products with little oversight or restriction. It's a bit like the Wild West—lawless and untamed.

Even decades after the 1976 Toxic Substances Control Act was enacted, few of the approximately eighty-five thousand chemicals that have been produced for use in commercial products, many of which have been identified as potential threats to human health, have even been tested, let alone banned or regulated. In the rare instances when chemicals *are* tested, the studies that are conducted don't usually protect human health because the protocols don't address the effects of dosing nuances (high versus low, for example). Or, they don't consider the potentially cumulative or interactive effects these substances can have when they're mixed inside the human body.

The point is, myriad chemicals that are used to manufacture a vast array of consumer products are largely unregulated. Which means they continue to be on the market, and we continue to buy them and bring them into our homes, where they get into our bodies. Once they're on the market, these chemicals can enter our bodies in numerous ways—in the contaminated foods and beverages we ingest, in microscopic airborne particles we inhale, and in the products we absorb through our skin.

The Chemical Class Name Game

To understand how harmful chemicals linger in the environment, it helps to distinguish between persistent and nonpersistent chemicals. "Legacy chemicals" stick around and can cause problems long after they're released into our bodies and the environment. These include persistent organic pollutants (POPs) such as dioxin (a by-product of industrial processes), dichloro-diphenyl-trichloroethane (DDT, a pesticide), and polychlorinated biphenyls (PCBs, industrial compounds). The adage that "nothing lasts forever" isn't true of these chemicals, which were designed precisely to last; they remain in the environment and our bodies for years. The trouble is, these "forever chemicals" have the potential to do endless harm once they get into the bodies of humans and other species. Because they are not water-soluble, they don't degrade, and they are stored in body fat and other tissues.

The Stockholm Convention on Persistent Organic Pollutants, a global legally binding agreement adopted in 2004, outlaws the production, use, and release of all persistent organic pollutants. It listed twelve of the most toxic substances—aldrin, endrin, dieldrin, furan, hexachlorobenzene, PCBs, chlordane, DDT, dioxins, heptachlor, mirex, and toxaphene—as priorities for elimination. Despite the adoption of this international agreement, many countries, including the United States, have not ratified it, so use of some of these toxic chemicals continues. As a result of current and past use, these POPs continue to be found in our air, soil, water, and food—and in our bodies, as well as in the bodies of other species.

Once they enter the human body, from the foods we eat, the air we breathe, and the water we drink, these chemicals are stored in fat tissue, where they can accumulate and remain for years. DDT, for example, has a half-life in humans of up to fifteen years. (If you think that means it is gone after fifteen years, nope. That's how long it takes for its concentration to fall to half of its original value.)

By contrast, nonpersistent chemicals such as BPA, phenols, and phthalates are water-soluble, which means they essentially wash out of our bodies and the environment, and they do not accumulate in the body's fat. These short-lived chemicals have half-lives of four to twenty-four hours. Even so, levels of human exposure to many non-persistent chemicals—such as phthalates and phenols—tend to be fairly stable because of our continual use of products that contain them.

Chemicals are so pervasive in our modern world that it's impossible to avoid them entirely. We're exposed to these chemicals on a daily basis, often without realizing it. Many of these chemicals, particularly phthalates and flame retardants, are even present in household dust, small particles of which can be inhaled, ingested, or absorbed through the skin. Even if you lived in a hygienic bubble, there's a good chance that some of the materials used to make it would contain plasticizers, adhesives, or other chemical components that could have endocrine-disrupting effects.

Not every human being is equally affected, however. As Norah MacKendrick, PhD, an associate professor of sociology at Rutgers University, writes in *Better Safe Than Sorry*, "While all bodies contain synthetic chemicals, body burdens differ in crucial ways that reflect the social and political organization of risk, gender, and social inequalities." For example, while men and women are both exposed to these chemicals daily, most cosmetics—hair products, creams, lotions, et cetera—are primarily marketed to women, and these substances contain a cocktail of heavy metals and endocrine-disrupting chemicals. But for most other chemicals, including testosterone-lowering phthalates, men have a greater overall exposure.

Children are at risk, too, even before their first day of life. Babies are now entering the world already contaminated with chemicals because of the substances they absorb in the womb. And once the infants emerge, they consume many "forever chemicals" that are stored in the fat in their mother's breast milk. The longer the mom breastfeeds, the more she unloads, particularly for her firstborn child. In the 2010 Swedish documentary *Submission*, a Swedish actress who is pregnant has her blood tested for EDCs and is horrified by the results. An older woman chimes in, "I instantly thought of my sons and how long I nursed them." This is a particularly painful realization for women who believe that they're boosting their babies' immune function and brain development by breastfeeding.

Wreaking Hormonal Havoc

Once they're inside us, environmental toxins do their damage in a variety of ways. One of the sneakiest is through endocrine disruption, interfering with the body's endocrine (or hormone) system. Endocrine-disrupting chemicals (EDCs) can interfere with the normal function of the body's endocrine system, a complex network of glands and organs that produce and secrete hormones. As you've read, hormones are chemical substances that are produced in one part of the body, then travel, like messengers carrying important information, through the bloodstream to other parts of the body, in order to regulate how certain cells and organs fulfill their functions. Many different types of hormones are in the human body; given the subject of this book, I will focus primarily on the reproductive hormones, particularly estrogen and testosterone, which is the major androgen that stimulates the development of male characteristics.

Some EDCs act like impostor hormones and bind to receptor sites where the natural androgen or estrogen is supposed to dock, thereby fooling our bodies into responding to them as if they're the real deal. Sometimes this results in too much or too little of that natural hormone being produced or released; other times, this can alter the transport of hormones, changing where they go, which may thwart them from doing their assigned tasks. Other EDCs can affect how naturally occurring hormones are broken down or stored in the body, thereby increasing or decreasing the levels of these hormones in the bloodstream, and still other EDCs can alter our bodies' sensitivity to different hormones. When a synthetic external chemical changes the way a hormone is supposed to act inside the body, physical abnormalities can develop in cells and tissues, and an organ may not function the way it should. EDCs can have antiandrogenic properties or potent estrogenic properties; as you might expect, antiandrogens are particularly problematic for boys, while the estrogenic ones are worse for girls.

The breadth of the potentially disruptive influences of EDCs is striking. They have been linked to numerous adverse health effects in almost all biological systems, not just the reproductive system but also the immunological, neurological, metabolic, and cardiovascular systems. To make matters worse, an individual's genetic susceptibility to certain health conditions, coupled with exposures to other chemicals and lifestyle habits, can increase the effects produced by a particular EDC.

Endocrine-disrupting chemicals can also have profound effects on the developing brain in ways that can affect a person's gender and sexual identity. You may have heard that the brain is the most powerful sexual organ. Sex therapists often say this because the brain is what activates sexual arousal and responsiveness. Well, here's an interesting twist: In 2014, my colleague Bernie Weiss, PhD, who was then a toxicologist at the University of Rochester, spoke in a different way about the brain as the biggest sexual organ in the body. He was referring to how certain environmental chemicals can alter brain function and behavior with different impacts on males and females. It's not just what's between a person's legs that reflects his/her/their sex or gender; the brain does, too. Chemicals in our environment may influence not only the development of these sex-determining organs but also behaviors that are typically different in boys and girls. For example, boys tend to acquire spatial ability (the capacity to understand and remember the spatial relations among objects) earlier, while girls' language skills often develop at a younger age than boys' do. My research and that of others has shown that higher exposure to some hormone-influencing chemicals can decrease male-female differences in these kinds of abilities.

Once they're mobile, young children are particularly at risk of exposure to chemical-laden household dust because they crawl, play on the floor, and frequently put their hands in their mouths. Because their bodily systems are just developing, young children are less able to metabolize these chemicals than adults are. Even small exposures can add up. Once these chemicals enter our bodies, at any age, they can be widely distributed throughout various systems from head to toe. How far they can travel in our bodies is truly astonishing. (Cringe alert: in 2018, for the first time ever, microplastic particles—nine different types!—were found in human stool, among volunteers from Finland, the Netherlands, the United Kingdom, Italy, Poland, Russia, Japan, and Austria.)

If you don't think you're exposed to these chemicals regularly, consider this: While writing their book *Slow Death by Rubber Duck*, Canadian environmentalists Rick Smith and Bruce Lourie set up an experiment to examine how products that are commonly used in daily life alter the body's chemical burden, using themselves as subjects. In the summer of 2008, Rick had called and asked me to serve as the "phthalate expert" on their science experiment and review its protocol and results. Guided by the principle that their exposures had to mimic those of real life, Rick and Bruce focused on chemicals of concern and identified activities that were likely to increase their exposure to these chemicals. Before the experiment began, they determined their personal baselines by having concentrations of these chemicals measured in samples of their blood and urine.

They designed a "test room" in Bruce's condo and stayed there in twelve-hour shifts, exposing themselves to the test chemicals by applying personal-care products, using antibacterial hand soap, eating canned or packaged foods, drinking coffee or canned soda, and hanging out in the room where the carpet and the couch had just been protected with Stainmaster, which is designed to help materials resist stains. After four days, they collected more urine and blood samples and had them sent to a high-precision lab for analysis. While the levels of the test chemicals increased significantly, from baseline to four days later, there was one standout, as Rick noted in the book: "The really dramatic result was that as a result of my product use, my MEP [monoethyl phthalate] levels—one of the chemicals that Shanna Swan had connected with male reproductive problems—went through the roof, from 64 to 1,410 nanograms per milliliter." This was a direct result of smearing himself with scented toiletries, including hair-care products, shaving gel, deodorant, fragrance, and lotions, as well as using scented liquid soap and a plug-in scented oil in the test room.

Since 1999, the National Health and Nutrition Examination Survey (NHANES) has assessed the health of twenty-five hundred adults and children in changing representative population samples, and the Centers for Disease Control and Prevention has periodically measured the levels of environmental chemicals in these study participants. This research tells us about who is being exposed to which chemicals and when, which helps scientists map exposures and the associated risks across different populations; in other words, it allows us to find exposure hot spots and study them. This is important because while we can ask people how much they smoke or how much Tylenol they take to try to gauge the levels of these chemicals in their bodies, we can't do the same with environmental chemicals. After all, none of us knows exactly how much we're exposed to these chemicals or how much of them may be in our bodies, so asking such questions would be pointless. Instead, environmental chemists have developed methods to measure even low levels of chemicals in tiny amounts of body fluid, usually urine and blood, but also breast milk and others.*

Not surprisingly, the number of chemicals tested has increased over time, as new ones become more commonly used in commercial products

---

* The persistent chemicals that are stored in fat (such as DDT) are best measured in the blood, while the nonpersistent ones (such as phthalates) are most reliably measured in urine.

and/or raise concerns. For reproductive health, phthalates, bisphenol A, flame retardants, and pesticides are of paramount concern—with phthalates having the strongest influences on the male side of the equation, while BPA is a particularly bad actor on the female side. Given how quickly not only industry but also the public embraced "better living through chemistry," including plastic and other modern chemically based conveniences, it's not surprising that we saw a decline in sperm counts after the 1950s, a time when chemical production was rapidly increasing. Let's take a closer look at these chemical culprits' effects.

Phthalates

A large, diverse class of chemicals, phthalates are found in plastic and vinyl, floor and wall coverings, medical tubing and medical devices, and toys, as well as in a vast array of personal-care products (including nail polishes, perfumes, hair sprays, soaps, shampoos, and others). Phthalates are widely distributed throughout the body and can be measured in urine, blood, and breast milk. The most concerning phthalates are those that can decrease the production of male hormones such as testosterone (the antiandrogenic phthalates) that the male needs to become fully masculinized, changes that can make him more likely to be infertile or to simply have a lower sperm count. In this respect, the three particularly bad actors are di-2-ethylhexyl phthalate (DEHP), dibutyl phthalate (DBP), and butyl benzyl phthalate (BBzP). Because of their reproductive toxicity, these three phthalates are scheduled to be gradually phased out in the European Union, along with others; that's not the case in the United States, though.

Of these three notorious phthalates, DEHP appears to be the most damaging to the male reproductive system. A 2018 review of research on the subject found "robust evidence of an association between DEHP and DBP exposure and male reproductive outcomes," including shorter AGD, reduced semen quality, and lower testosterone levels with DEHP, and reduced semen quality and a longer time to achieving pregnancy with DBP. Men with high exposure to phthalates during adulthood also tend to have lower sperm counts and more abnormally shaped sperm.

As you saw in chapter 5, prenatal exposure to antiandrogenic phthalates can alter male reproductive development in the infant, including the size of the genitals. Preliminary data suggest that by early adulthood men whose mothers had higher concentrations of several phthalates during pregnancy have reduced testicular volume, which is associated with lower testicular function (including worse sperm parameters). It's an unfortunate cluster of effects, from multiple perspectives. Studies have shown that young men with higher levels of phthalate metabolites—which are by-products of metabolizing the chemical in our bodies—have poorer sperm motility and morphology. This is bad news, since higher levels of phthalate metabolites also are associated with increased sperm apoptosis—a term for what is essentially cellular suicide. It's safe to assume that no man wants to hear that his sperm are self-destructing.

Phthalates are bad news for women's ovaries, too. High levels of phthalate exposure have been linked with anovulation (when ovaries don't release an egg during a menstrual cycle) and polycystic ovary syndrome (PCOS), a hormonal disorder involving abnormal ovarian function and elevated levels of androgens. Moreover, there's some evidence that higher blood levels of metabolites of certain phthalates may be linked with primary ovarian insufficiency (aka premature ovarian failure). In addition to potentially moving up the timing of menopause, it appears that heavy exposure to phthalates from personal-care products, in particular, is associated with a greater frequency of hot flashes in women ages forty-five to fifty-four. Yet most women don't realize that their grooming practices may come with this hidden cost to their well-being at midlife.

In 2002, a coalition of environmental and public health organizations tested seventy-two name-brand beauty products for the presence of phthalates and found that nearly three-quarters of the products, including deodorants, fragrances, hair gels and mousses, and hand and body lotions, contained these chemicals. In 2004, the European Union banned the use of DEHP and DBP in cosmetics; while the United States hasn't followed suit, some companies have voluntarily decided to phase out their use in personal-care products. That's at least a step in the right direction.

Bisphenol A

BPA was first synthesized in 1891, but not until the period between the two world wars were its commercial possibilities explored. In the mid-1930s, British medical researcher Edward Charles Dodds, at the University of London, identified the estrogenic properties of BPA, and for the next several years he continued testing chemical compounds as he searched for a powerful synthetic estrogen. He found it in diethylstilbestrol, better known as DES, estimated to be five times more potent than estradiol, the most powerful estrogen that occurs naturally in mammals. Starting in the 1940s, DES was used for a range of "therapeutic" purposes, including those related to menstruation and menopause. The most dangerous use, by pregnant women to prevent miscarriage, wasn't banned until 1971, when it was discovered that it caused a rare cancer in the women's daughters.

Though it has a similar chemical structure to DES, BPA was never used for pharmaceutical purposes. Instead, its utility was found to be in plastics. Starting in the early 1950s, BPA was used in epoxy resins that were incorporated into protective coatings on metal equipment, piping, and the lining of food cans, as well as into adhesives, nonskid coatings, and plastics. Over time, BPA started being used in hard plastics, electronics, safety equipment, thermal receipt paper, and other everyday items—until it became ubiquitous, despite that its estrogen-like properties continued to lurk in the background. Over time, it has been discovered that BPA exposure—particularly occupational exposure—is related to decreased sperm quality in men. When researchers from Kaiser Permanente conducted a study with factory workers in China to evaluate the effects of exposure to BPA, they found that men with detectable levels of BPA in their urine were more than four times as likely to have lower sperm counts, more than three times as likely to have poorer sperm vitality, and more than twice as likely to have lower sperm motility than those with undetectable BPA in their urine.

There can be other damaging ripple effects. The sons of men with high BPA exposure often have a shorter AGD (the span from the anus to the base of the penis). And when researchers examined sexual satisfaction in men who worked in factories that manufactured BPA and epoxy resin, they found that these men had higher rates of sexual dysfunction, including more erectile dysfunction and ejaculation difficulty and decreased sexual desire.

The potential effects on women's reproductive health are even greater, in part because by mimicking the female hormone estrogen, BPA can induce estrogen-like changes in the body. There's compelling evidence that women who have high blood levels of BPA may have an increased risk of fertility challenges, including difficulty becoming pregnant; whether this is because the chemical has a detrimental effect on the function of various reproductive organs, or on the proper cycling of estrogen levels, which is crucial for ovulation, isn't clear.

Among women who do get pregnant, those who have the highest levels of conjugated BPA in their blood have an 83 percent increased risk of miscarriage during the first trimester. Women who have higher BPA concentrations in their urine during the first trimester of pregnancy are likely to give birth to daughters with a significantly shorter AGD. BPA is also believed to contribute to polycystic ovary syndrome (or PCOS), given that studies in humans have found that blood concentrations of BPA are higher in women with PCOS than in "reproductively healthy women." In addition, exposure to BPA during early life and adulthood has been correlated with poor egg quality and named as a possible culprit in premature ovarian insufficiency, leading to an earlier age of menopause. Throughout a woman's life, BPA might as well be considered a nemesis to her reproductive health.

Flame Retardants

Since the 1970s, chemical flame retardants have been added to numerous materials to prevent or slow the growth of fire, in foam and upholstered furniture, mattresses, carpets, children's pajamas, computers, and other common products. There are dozens upon dozens of different flame retardants. While some have been removed from the market due to health or safety concerns, these gone-but-not-forgotten chemicals don't break down easily; rather, they persist in the environment and can build up in fatty tissues in humans and animals. (The latter means we ingest these chemicals from the animal fat we consume.)

Over the years, flame-retardant chemicals have been found to have adverse effects on human health. A class called polybrominated diphenyl ethers (PBDEs) is associated with neurodevelopmental problems in children and altered thyroid function in pregnant women. These chemicals also exhibit a range of endocrine-disrupting activities, from estrogenic action to anti-estrogenic properties to antiandrogenic activity. Given these effects, it's not surprising that research has found that it takes longer for women with higher PBDE concentrations in their blood to get pregnant. The risks don't end once a woman gets pregnant, though, because there's also evidence that high blood levels of these chemicals are associated with an increased risk of miscarriage.

Meanwhile, prenatal exposure to high levels of PBDEs can alter the timing of the offspring's puberty, most notably leading to a later onset of menstruation in girls but early puberty in boys. When a developing fetus is exposed to PBDEs and other brominated flame retardants in utero, these chemicals can have disruptive effects on the fetus's endo-crine system, primarily on thyroid function, but also on reproductive function and neurodevelopment. Evidence is also mounting that these chemicals, like many others, can build up in human breast milk and be transferred to babies who are nursing. In a study published in 2017 researchers examined PBDE concentrations in human breast milk collected in North America, Europe, and Asia over a fifteen-year period: total PBDE concentrations were more than twenty times higher in breast milk in North America than in Europe or Asia. *So much for the purity of mother's milk!*

Pesticides

Pesticides—including herbicides, insecticides, and fungicides—also can have adverse effects on human health, including our reproductive potential and endocrine systems. Depending on the chemical agent, these effects can include competitive binding to estrogen, progester-one, or androgen receptors. Alternatively, they can inhibit androgen or estrogen production, availability, or action—or potentially increase the production of female hormones such as estrogen or progesterone. Still others can cause disruptions in thyroid hormone production or action. It's a bit of a free-for-all.

In the summer of 1977, a small group of pesticide-production workers in Lathrop, California, were worried about how the chemicals were affecting their health. As one worker at the Occidental Chemi-cal plant recalled, "It was rumored [that] anybody that worked in that department for more than two years couldn't produce children. And I haven't." Soon the results of testing revealed substance behind these rumors: many workers on the production line were found to have abnor-mally low sperm counts, as little as zero in some cases. Their sterility was eventually linked to their exposure to dibromochloropropane (DBCP), which had been widely used on pineapple and banana plantations and was once the most heavily used pesticide in the United States, until it was banned from use in 1979.

Soon after that, workers who had long-term exposure to ethylene dibromide (EDB) from treating fruit-fly infestation in papayas in Hawaii were found to have significant decreases in sperm quality compared to workers from a nearby sugar refinery.

In South Africa, the insecticide DDT is still widely used in an effort to control malaria. In addition to having detrimental effects on the reproductive development of various forms of wildlife, researchers found that DDT exposure was associated with impaired semen quality and external urogenital birth defects in males born to mothers whose houses were sprayed; they also found that adult men living in villages where the houses were routinely sprayed with this endocrine-disrupting chemical have higher estrogen *and* testosterone concentrations.

In 2000, I launched the Study for Future Families, which examined semen quality in men recruited from four very different parts of the country. We found that the most dramatic differences in these reproductive parameters were between men from rural central Missouri and urban Minneapolis. The Minnesota men had twice as many moving sperm as those in central Missouri, which had far greater quantities of farmland and pesticide use. To test the possibility that pesticide exposure could be to blame, my colleagues and I selected a group of men in whom all sperm parameters were low and a group of their peers who had high values for all sperm parameters, then measured pesticides in their urine. You can probably guess the results—the Missouri men had been exposed to several herbicides and insecticides and had worse sperm quality.

Pesticide exposure can also occur when people consume pesticide-contaminated foods, but it's not clear to what extent this can affect reproductive health in men. In a 2015 study from Spain, researchers examined urinary concentrations of certain pesticide metabolites in men at an infertility clinic and found that sperm concentration and total sperm count were lower in men with higher concentrations of four different pesticide by-products in their urine. There also was a significant adverse association between the percentage of motile sperm and metabolite concentrations of three different pesticides in their urine.

Women don't get a free pass when it comes to pesticides, either. In a study involving 1,710 pregnant women and their male spouses in Greenland, Ukraine, and Poland, researchers examined the women's blood samples for the presence of certain pesticides and whether they had a history of miscarriage or stillbirth. Women who had higher blood levels of two pesticides—one of the PCBs (CB-153) and DDE (a metabolite of DDT)—had a significantly higher risk of pregnancy loss. Some scientific evidence also suggests that it may take longer for women who have high exposure to organochlorine pesticides to get pregnant.

These findings don't just apply to farmworkers; to some degree, depending on a particular pesticide's toxicity and the person's level of exposure, exterminators, gardeners, greenhouse workers, and florists could also be at risk. So could people who consume, usually without even realizing it, a high volume of foods and beverages that have pesticide residues.

Other Under-the-Radar EDCs

The hidden hormonal threats don't stop there. Higher levels of perfluoroalkyl compounds (PFCs)—which are stain-, water-, and grease-repellent chemicals found in a wide range of consumer products, including fast-food packaging, paper plates, stain-resistant carpets, and cleaning solutions—in men's blood and semen are correlated with a reduction of semen quality, testicular volume, penile length, and anogenital distance. Some evidence suggests that women with a moderate to high exposure to PCBs (polychlorinated biphenyls) from eating contaminated fish are susceptible to shortened menstrual cycles and reduced fecundity. (Despite being banned in the United States, PCBs persist in the environment and accumulate in the food chain.)

In a noteworthy study, Russian boys who were found to have high blood concentrations of certain dioxins, which are by-products from industrial practices that persist in the environment, at age eight or nine had lower sperm counts, concentrations, and motile sperm counts at age eighteen or nineteen. Dioxin can adversely affect a woman's reproductive health, too. An explosion in 1976 at a chemical factory near Seveso, Italy, led to the highest-known population exposure to a dioxin called TCDD, which is short for 2,3,7,8-tetrachlorodibenzo-p-dioxin. Researchers measured blood levels of TCDD among 601 women ages thirty and younger and tracked their health over twenty years; those who had high blood concentrations of TCDD had double the risk of endometriosis as their peers who had lower levels. In addition, high blood levels of TCDD were associated with a longer time to pregnancy and double the risk of infertility.

If it sounds like we're living in an alphabet soup of evil chemicals, well, we are. And this list doesn't even include the pharmaceuticals we're exposed to!*

One more kicker: Contrary to the widely held assumption that "the dose makes the poison," which was based on the notion that only a high enough concentration of a toxic substance could cause harm, endocrine-disrupting chemicals often don't behave this way. Rather, they can have harmful impacts even at very low doses. These low doses occur not from occupational exposures or industrial accidents, but with ordinary, every-day contact such as simply putting on makeup or body lotion or even carrying this book around in a plastic bag.

---

*   Pharmaceutical drugs are likely lurking in our water supply because currently most municipal water-treatment facilities are unable to remove them from drinking water. This means that we are consuming trace amounts of pharmaceutical agents, including analgesics, antibiotics, anticoagulants, antidepressants, antihistamines, antihypertensive drugs, hormones (from oral contraceptives and hormone therapy), and muscle relaxants, in our tap water. In addition, chemicals from personal-care products, such as shampoos, conditioners, body washes, and lotions, are also running down the drain and into water-treatment plants; their chemical ingredients are not all filtered out before they reach your tap. Which means this is yet another way EDCs can get into your body.

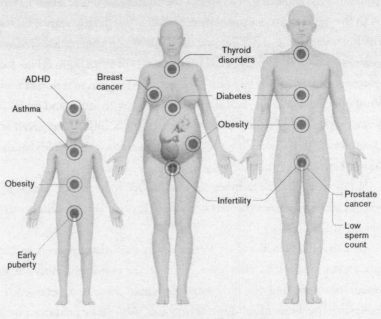

Endocrine-Disrupting Chemicals: Low Doses Matter

Everyday exposures contribute to modern health epidemics.

Thyroid disorders

ADHD

Breast cancer

Diabetes

Asthma

Obesity

Obesity

Infertility

Prostate cancer

Low sperm count

Early puberty

How are people exposed?

Children's toys, plastic drinking bottles, cleaning supplies, house dust, home furniture/electronics, building materials, fragrances, food, food packaging, thermal cash register receipts, drinking water, personal-care products

SOURCE: HEALTH AND ENVIRONMENTAL ALLIANCE AND TED X

Regrettable Substitutions

It would be nice to think that when a particular chemical is found to be harmful and others are substituted for it during the manufacturing process, the problem is solved. But sadly it doesn't always work out that way, since the chemicals that are substituted can have the same effects as the chemicals they are replacing. This pattern played out in the 1970s when DDT was thought to be a "safe" replacement for the pesticide lead arsenate, which was found to be neurotoxic. When DDT also was found to be neurotoxic, it was replaced with organophosphate pesticides, another class that also has neurotoxic effects that interfere with a child's brain development.

In my own studies, we saw this as well. During the ten years (2000 to 2010) between recruitment for our two large studies of pregnant women, people's exposure to di-2-ethylhexyl phthalate (DEHP), a chemical used as a plasticizer, had declined 50 percent, due in part to its ban in children's toys. Without question, the ban was a good thing for public health and environmental health—except that in the meantime the DEHP was replaced by chemical substitutes, including diisononyl phthalate (DINP), which turned out to be as damaging to male reproductive development as DEHP.

Similarly, while PBDEs were banned in 2004, one of the chemicals used to replace them has turned out to be nearly as dangerous. When it was released by Dow Chemical Company in 2011, Polymeric FR, which is used mostly behind roofs and walls, was touted as being an example of "breakthrough sustainable chemistry," but it turns out that its breakdown compounds look very much like old flame retardants—toxic. Another example: since bisphenol S was substituted for bisphenol A in many products touted as being "BPA-free," it has become apparent that these products also may interfere with endocrine function in ways that could promote premature puberty, obesity, and damage to a woman's eggs. *I'm sure you get the picture.*

The trouble is, there is nothing to stop "regrettable substitution," a practice in which manufacturers replace a harmful chemical with another chemical that may turn out *not* to be a safe alternative. This switcheroo can happen when industries respond to public outcry or regulatory pressures about a chemical's potential health effects by replacing one chemical that has been identified as harmful with a new one the public assumes is safe.* But that doesn't always turn out to be true.

---

*   In essence, this takes "advantage of the public's misperception that the re-

As Ruthann Rudel, MS, a toxicologist at the Silent Spring Institute, a research center in Newton, Massachusetts, told a writer for the *New York Times*, "Sometimes we environmental scientists think we are playing a big game of Whac-A-Mole with the chemical companies." It may be a fun game for kids, but we shouldn't be playing it with our reproductive health.

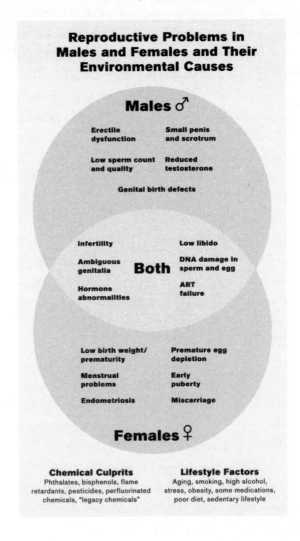

Reproductive Problems in Males and Females and Their Environmental Causes

Males ♂

Erectile dysfunction

Small penis and scrotum

Low sperm count and quality

Reduced testosterone

Genital birth defects

Both

infertility

Low libido

Ambiguous genitalia

DNA damage in sperm and egg

Hormone abnormalities

ART failure

Low birth weight/ prematurity

Premature egg depletion

Menstrual problems

Early puberty

Endometriosis

Miscarriage

Females ♀

Chemical Culprits
Phthalates, bisphenols, flame retardants, pesticides, perfluorinated chemicals, "legacy chemicals"

Lifestyle Factors
Aging, smoking, high alcohol, stress, obesity, some medications, poor diet, sedentary lifestyle

placement is inherently safe," as the Collaborative on Health and the Environment notes.

Part III

The Reverberating Fallout

8

THE LONG REACH OF EXPOSURES:
Reproductive Ripple Effects

Health Spirals

It would be naive to think that the effects of fertility challenges or reproductive anomalies would stay in their lane, without having other consequences. These difficulties can affect a person's sex life, his or her ability to conceive the old-fashioned way, the person's self-image and body esteem, and his or her sexual relationship and emotional state. But the ripple effects don't stop there. Low sperm count, recurrent miscarriages, and reproductive disorders, such as endometriosis and polycystic ovary syndrome (PCOS), can have profound repercussions for a man's or woman's long-term health and even lead to premature mortality.

Let's start with guys. One thing many people don't realize is that compromised reproductive health, including low sperm concentration and low testosterone levels, is associated with decreased overall health among men. In a 2016 study of some thirteen thousand men who'd been diagnosed with male factor infertility, researchers found that men with low sperm concentrations had a 30 percent increased risk of developing diabetes and a 48 percent increased risk of developing ischemic heart disease, compared to men without the infertility diagnosis. Male infertility, including low sperm concentration, is also associated with an increased cancer risk, particularly testicular cancer and high-grade prostate cancer. Men with sperm concentrations below 15 million/mL had a 50 percent greater risk of being hospitalized for *any medical reason* than those with sperm concentrations above 40 million/mL, according to one 2017 study.

Given these elevated risks, it's not surprising that men with infertility can expect to die earlier than their more fertile peers. In a 2014 study, researchers from Stanford University followed the health of twelve thousand men who were evaluated for infertility and found that those with impaired sperm counts, sperm motility, or semen volume—any of which qualify as male factor infertility—had higher mortality rates during the subsequent decade than those with normal semen quality. Men with two or more abnormal semen parameters, which the researchers consider "severely impaired" semen, had a 2.3-fold higher risk of dying in the ten-year follow-up period than those with normal semen quality.

The exact mechanisms behind these links aren't known, but there are theories about what could be going on. One suggests that defects in DNA repair mechanisms impair cellular division processes in ways that affect sperm production and increase the likelihood of cancer developing. Another theory points to a hormonal explanation, namely that men with infertility have lower circulating testosterone levels than fertile men do; low testosterone levels in men can increase their risk of developing cardiovascular disease and set the stage for muscle loss, increased abdominal fat, weakened bones, erectile dysfunction, and memory, mood, and energy problems, circumstances that many men desperately want to avoid. Researchers also hypothesize that in utero disruptions in genetic programming can impair not only genital development but can also affect the man's health later in life. It's a tangled web of contributing factors, indeed.

Whether it's called "a sixth vital sign," a "harbinger," or "a fundamental biomarker," this much is clear: a man's semen quality can tell him something about his future health risks. On the upside, men with high-quality semen have a longer life expectancy and a decreased incidence of a wide range of diseases compared to their peers with infertility, according to a study of forty thousand Danish men who were followed for up to forty years. Simply put, having an abundant sperm supply is associated with better health for men—virility on multiple fronts.

Unfortunate Domino Effects for Women

For women, there are also strong associations between reproductive health and future well-being. Those with PCOS often have insulin resistance or diabetes and suffer from metabolic syndrome, which increases their risk for developing cardiovascular disease in addition to decreasing fertility. Women who start menstruating early (before age twelve) may have a 23 percent higher risk of dying young from any cause than their peers whose first periods arrive later, probably because early puberty in girls is associated with an increased risk of developing obesity, type 2 diabetes, asthma, and breast cancer. Anovulation, which is the failure of the ovary to release an egg during a given menstrual cycle, has been linked with an increased risk of uterine cancer, while endometriosis and tubal factor infertility could increase a woman's risk for ovarian cancer.

Women who are diagnosed with infertility are also at higher risk for hormone-sensitive cancers. This makes sense given that they don't get a break from the hormonal ups and downs associated with menstruation: pregnancy offers women a nine-month hiatus from their periods, plus another month or so after giving birth if they're *not* breastfeeding and up to six months or longer if they are breastfeeding exclusively. This is significant because uninterrupted menstrual cycles (as in never getting pregnant) means nonstop exposure to ovarian hormonal fluctuations, which stimulate cell growth in the breasts, ovaries, and endometrium. To a lesser extent this is also true of women who have their first child late in life (or never have a child). A woman who has her first child at age forty or later has a fourfold greater risk of developing breast cancer compared to a woman who has a child at age fifteen, largely because the older women have gone decades without getting an extended holiday from hormonal stimulation.

In a 2019 study involving more than sixty-four thousand women who were diagnosed with or tested or treated for infertility and more than three million women who were seen for routine gynecological care, researchers from Stanford University sought to investigate whether similar risks apply to other cancers by tracking the women's health over several years. It turned out that the women who were seen for infertility testing and treatment had an 18 percent higher risk of developing uterine, ovarian, thyroid, liver, and pancreatic cancers, as well as leukemia. Interestingly, among the women who were classified as infertile yet became pregnant and gave birth to a child during the follow-up period, the risks of uterine and ovarian cancers dropped down to match those of their naturally fertile counterparts.

In addition, the lifestyle and chemical-related stressors that can alter a man's or woman's reproductive health can also modify the expression of his or her genetic code and possibly affect future generations of their families.

Tinkering with the Master Plan

How might these hand-me-down effects work? That's the domain of a field called epigenetics, which literally means "on top of genetics." The term, coined by British scientist Conrad Waddington in 1942, refers to the study of biological mechanisms that can change gene function and expression, for example, by switching particular genes on and off, or dialing their expression up or down, without altering the underlying sequence of DNA. In the span of several decades, the field has blossomed and provided new insights into how a person's environment, including his or her exposure to certain chemicals and lifestyle practices, can influence the expression of certain genes, which can then alter his or her risk of developing specific health conditions.

Here's where things get complicated. Some scientists use the word *epigenetics* to refer to chemical or physical changes that affect gene regulation without altering the essential DNA sequence. By contrast, others believe the term should apply only to changes that are heritable—meaning, passed from one cell to another or from one organism to another. If you're finding it difficult to wrap your mind around all this, you're in good company. As Siddhartha Mukherjee, MD, noted in *The Gene: An Intimate History*, "The shifting meaning of the word *epigenetics* has created enormous confusion within the field."

Here's the gist of what you need to know: Your genes and your environment can interact in ways that can change how your genes are used or expressed. That alone is amazing enough, but here's the really astounding part. The food we eat, the air we breathe, the products we use, and the emotions we feel have the potential to influence not only how our own genes are expressed but also how those of our unborn descendants might behave in the future. That's right—our lifestyles and environments can have ripple effects on the health and development of our unborn children and grandchildren through mechanisms that foster cellular memory and can be maintained across several generations.

Such effects are considered transgenerational when they're seen in a generation that was *not* directly exposed to the stimulus in question, such as the sons and daughters of a parent who *was* exposed. When the effects extend to the second, third, or fourth generation after the generation that had the initial exposure, then they're deemed to be multigenerational. Together, these passed-along influences can be considered *intergenerational*, an all-encompassing term I prefer for simplicity's sake.

An analogy: Imagine that a documentary was being made about the development and maturation of your body. The genes you carry would provide the script, outlining key actions or events that would be featured in the film; epigenetic changes would reflect alterations or tweaks a director might make to how the script is performed—in this case, by causing certain sets of genes to be turned on (expressed) or turned off (inhibited or silenced). In other words, the director (epigenetic changes) has the power to yell "Action!" or "Cut!" or to suggest putting a different spin on a particular event.

In real life, epigenetic changes, which are part of the normal development, health, and survival of the species, can influence a person's risk of disease throughout the life span. When someone is exposed to a particular stimulus— whether it's a toxic chemical, intense stress, or a certain dietary factor—this influence can elicit epigenetic modifications that can have lasting effects on the person's development, metabolism, and health—and sometimes even the development and health of that person's offspring.

We know of three primary epigenetic mechanisms, which I'll explain here, and others will probably be identified down the road. One of the best characterized is DNA methylation, a chemical process that adds a methyl group (a common structural unit of organic compounds) to DNA. DNA methylation, which helps regulate major cellular processes, essentially acts like a switch that dials the activity of genes up or down by modifying a gene's interactions with the machinery within a cell's nucleus. In another epigenetic mechanism, histones, which are proteins that serve as a spool around which DNA is wrapped, can be modified through specific chemical processes; a particular histone modification can then precisely calibrate gene expression.

A third epigenetic mechanism involves RNA (short for *ribonucleic acid*), which is present in all living cells and plays essential roles in the coding, regulation, and expression of genes. The RNA-silencing mechanism is a modification during which the expression of one or more genes is down-regulated or suppressed by small noncoding stretches of RNA. Without getting into the weeds on noncoding RNA function, suffice it to say that these RNA molecules can alter gene expression and play a key role in biological processes. One way or another, all of these epigenetic mechanisms act as switches, modulators, or tags (which serve as a kind of cellular memory) that can change the epigenetic landscape. These changes are akin to editing and rewriting the script for your life's story.

Now, imagine that someone is using different-colored highlighters to mark up different parts of that script, to indicate which parts need to be read most carefully (say, orange) and which ones aren't as important (say, blue). The color-coding system can change throughout your lifetime, in response to environmental influences, so that a part that was once blue becomes orange or vice versa. In addition, some lines or stage directions can be passed down to your next of kin, just as some highlighted portions of a document still show up, as either a color or a shade, when it is photo-copied. This is the essence of how epigenetics works.

Undesirable Legacies

Your life story may not stop with you, however, and that may be the most astounding part. These epigenetic effects can influence a child's risk of developing asthma or allergies, obesity, heart or kidney disease, some neurological disorders, and some reproductive abnormalities. It has long been recognized that there's an intergenerational transmission—from mothers to children—of exposure to chemicals, metals, pharma-ceuticals, stress and trauma, and other detrimental factors, which makes sense intuitively since a mother's body is a baby's first home. Increasingly, research is suggesting that the same is true of men.

Here's one area where this has been illustrated: A parent's experience with war, trauma, or severe stress can have hand-me-down effects on the mental health of his or her offspring, even if the children don't grow up hearing stories of these horrors. The descendants of trauma survivors seem to inherit a biological memory of the hardship their parents endured—namely, through alterations in certain genes and levels of circulating stress hormones, according to Rachel Yehuda, PhD, a professor of psychiatry and neuroscience at the Icahn School of Medicine at Mount Sinai.

In one study, Yehuda and her colleagues interviewed adults with at least one parent who was a Holocaust survivor and adults whose par-ents hadn't been exposed to the Holocaust or experienced post-traumatic stress disorder (PTSD). Then, they took blood samples from the par-ticipants to compare methylation of a gene that's involved in the stress response (GR-1^F) and cortisol levels in response to being given low-dose dexamethasone (an anti-inflammatory drug). They found that subjects whose parents had experienced PTSD had alterations in methylation of this particular gene, a sign of trauma-induced epigenetic modifications.

This may sound like a lot of technical gobbledygook, but these alterations can have significant consequences for later generations. Another of Yehuda's studies, involving Holocaust offspring, found that a mother's PTSD significantly enhanced her child's risk for developing PTSD, while a father's PTSD significantly elevated his son's or daughter's risk for depression. Whether these effects are ultimately due to the parents' unpredictable behavior or epigenetic changes in a father's sperm has yet to be determined. But the idea that traumatic experiences can affect DNA in ways that are transmitted to subsequent generations, like molecular scars, is an upsetting family legacy.[*]

On the male side of the family tree, research involving mice found that the offspring of males who had significant stress before breeding displayed substantial alterations in stress reactivity of the hypothalamus-pituitary-adrenal (HPA) axis, which controls a person's or animal's response to stress—in this case, due to epigenetic reprogramming. This is especially noteworthy because altered stress reactivity is a prominent feature of PTSD, so it seems that a father's prior PTSD can become his child's PTSD through inherited molecular mechanisms. Taken together, these studies lend credence to the theory that psychological trauma or extreme stress can induce epigenetic changes that can be passed down from either parent or both and have real-life consequences for their children.

The Multigenerational Digestive Track

Another example of the intergenerational transmission of health effects: large fluctuations in the availability of food—from too little to plenty or the other way around—during grandparents' early years can have surprising trickle-down effects on subsequent generations. Research from Swe-

---

[*]   Yehuda's research does have its critics, who question how it's possible to separate the effects of children hearing horrific stories of the Holocaust from the influence of epigenetics. Another possible wrinkle: a chicken-or-egg question of whether DNA methylation is a result of trauma or whether DNA methylation increases the risk of experiencing PTSD.

den found that if a paternal grandmother experienced drastic changes in her access to food, from one year to the next up until puberty, her sons' daughters (her granddaughters) had a two and a half times higher risk of dying from cardiovascular disease as an adult. Similarly, babies exposed to nutritional deficiencies while in the womb during a severe famine in the Netherlands, known as the Dutch Famine (or Hunger Winter) of 1944–45, were found to be at increased risk of becoming obese and of developing schizophrenia as adults. (By contrast, women who were children ages two to six and experienced severe hunger during the Dutch Famine have been found to experience natural menopause earlier, compared to their peers who weren't exposed to the famine.)*

Dads aren't off the hook when it comes to diet. Studies have found that children conceived by undernourished fathers were heavier, and in some instances more obese, than children born to fathers and mothers who were well nourished before conception. By age nine, the sons of fathers who started smoking before age eleven have a higher risk of becoming overweight or obese; interestingly, while the sons of fathers who took up the smoking habit at an early age have a higher body mass index, the same isn't true of the fathers' daughters. Another body of research in male mice suggests that the sons of fathers who have a folic acid deficiency or who have the highest dose of folic acid supplementation have lower sperm counts. In other words, on the dad's side of the equation, there's no question that these paternal-lineage effects are transmitted via sperm.

Parental Advisories

Unsurprisingly, given all this, the lifestyle factors and environmental chemicals that a man or woman is exposed to can have reverberating effects on the reproductive health of generations to come. None of these

---

* This suggests that dramatic calorie restriction, from any cause, lowers the age at which a woman naturally goes through menopause.

potential epigenetic influences have slam-dunk effects, however: They don't occur in every single child who is born to parents who've experienced particular exposures that can induce epigenetic changes. But they *can* theoretically occur in any given child, and exposures in past generations make it more likely that these changes will happen.

That said, when these epigenetic changes do occur, just how many generations are affected by these exposures remains a matter of ongoing debate and research. It isn't clear, for example, whether the damaging effects of a particular exposure are perpetuated in both male and female children as well as third or fourth generations of descendants. The answer seems to depend on the culprit in question.

As an example, let's take a look at DES, which you'll recall was prescribed to millions of pregnant women until the 1970s because it was believed to prevent miscarriages. First of all, the treatment didn't prevent miscarriages—in fact, it increased the risk. Worse, it increased the incidence of certain reproductive disorders in the male and female offspring who were exposed to the drug in the womb. Most of the research on prenatal exposure to DES has focused on the reproductive effects in girls and women, and there are plenty, as you've seen.

Less well known are the potential effects in the boys and men whose mothers took DES during pregnancy, and these are significant. Not only can in utero exposure to DES increase the male babies' risk of having undescended testicles, hypospadias (misplaced urethral openings), epididymal cysts, and infection or inflammation of the testicles, but these boys also have higher chances of having micropenises (abnormally small but normally structured penises).* Whether there's also an association with decreased sperm counts or testicular cancer isn't clear because research on the effects of DES in sons hasn't been extensive.

---

* In case you were wondering, the definition of an "abnormally small penis" is not in the eyes of the beholder. It's a medical diagnosis that's made when the length of a penis is 2.5 standard deviations or more below the mean length. For an adult man, the average stretched penis length is 13.3 centimeters, or 5.2 inches, so a micropenis would be at most 9.3 centimeters, or 3.7 inches—a substantial shrinkage, indeed!

The real surprise: Some evidence suggests that the *sons* of females who were exposed to DES while in the womb—the *grandsons* of the expectant mothers who were exposed to DES—have an increased incidence of two genital abnormalities: undescended testicles and an abnormally small penis. In these instances, the DES damage can trickle down two or three generations, an effect that could be the result of epigenetic changes that are transmitted to subsequent generations through men.

Here's an example of how these effects can unfold with current chemical exposures and reproductive development. In a 2017 study, researchers examined phthalate levels in the urine of men undergoing in vitro fertilization (IVF) and found that several of these phthalates were associated with changes in sperm DNA (through what's called DNA methylation) that resulted in poorer embryo quality and a lower chance of successful implantation. The phthalates affected the genes that can influence a male baby's reproductive development and eventually a grown man's semen quality and fertility status—that is to say, whether or not *he* can have children. There's also evidence that a male's exposure to endocrine-disrupting chemicals can travel farther down the family tree, affecting reproductive development in successive generations of males. On the female side, research also has found that exposure to environmental toxins can lead to the intergenerational inheritance of PCOS or a premature reduction in the pool of viable eggs (aka diminished ovarian reserve).

Unfortunately, it gets worse because it appears that given the increasing number and volume of endocrine-disrupting chemicals and other toxins in our world, the damaging effects could be additive over time in descendants of the originally exposed person. In studies involving male mice, researchers at Washington State University sought to investigate the potential for this augmenting effect. So they tested the cumulative effects of the male mouse's prenatal and then his postnatal exposure to estrogenic chemicals, not just in one generation but also in three successive generations, and compared the severity of the effects across the various generations. They found that exposure to the endocrine-disrupting chemicals affected both the reproductive tract development and sperm production in the male mice that were exposed. No surprise there.

More startling was the finding that when subsequent generations were exposed to these endocrine-disrupting chemicals, the effects of the originally reported changes in sperm-producing cells were amplified. In addition, the incidence and severity of reproductive tract abnormalities—such as kinking or collapse of the vas deferens (which conveys sperm from the testicles to the urethra) and testicular fibrosis (which can lead to male infertility)—were increasingly observed, suggesting an additive effect. The impacts were worse in the second generation compared to the first generation and worse still in the third. That the damage got worse and worse as more generations were exposed suggests that male sensitivity to environmental estrogens is increased in successive generations that are exposed to common endocrine-disrupting chemicals, leading to a progressive decrease in sperm counts over multiple generations—a phenomenon that environmental scientist Pete Myers refers to as a "male fertility death spiral." This may sound like nothing more than a doomsday-themed video game or movie, but the possibility that the damage is getting worse and worse as subsequent generations are exposed to EDCs is beyond frightening. Where will the harm end?

Revising Our Reproductive Programs

Epigenetic and intergenerational effects such as these are significant and worrisome for humans and animals alike. After all, the evidence suggests that once these changes occur, the revised program for the future development of cells and body systems in successive generations could become permanent. It's as if the new pattern becomes etched in stone and cannot be altered or erased for either that particular man or possibly his future male heirs.

These findings shed light on the big revelation of my own research. That male sensitivity to environmental hormones is increased in successive generations of exposure—from a father to a son to a grandson—might explain the continued decline in sperm counts we saw over subsequent generations. As these second-, third-, and fourth-generation relatives are subjected to these harmful environmental influences, they become more sensitive to their effects, and there could be inherited DNA damage as well, which is yet another additive factor that can create a vicious cycle. Where in a family's lineage these damaging effects will stop, nobody knows.

There is, however, a glimmer of hope that some epigenetic effects may be reversible. For example, it's theoretically plausible that the propensity to become obese could be altered by changing the environment in the womb and the person's lifestyle in adulthood. Research in mice found that dietary supplementation with folic acid or genistein during pregnancy negates DNA hypomethylation and can counteract the damaging effects on the unborn pup of exposure to bisphenol A, an industrial chemical that's used to harden plastic (think baby bottles). That's the biological equivalent of clicking the "undo" function on your computer and erasing the error you just made.

But it isn't yet known to what extent future generations of human beings could be rescued from undesirable epigenetic changes or which effects are potentially reversible. Whether someone is lucky enough to escape this epigenetic cascade of unwanted intergenerational influences appears to be a matter of chance. Reproductive abnormalities, fertility challenges, and a higher risk of chronic diseases aren't acquired traits that any parent wants to pass along to his or her children. But our modern world has made it increasingly difficult to avoid these risks. That's why scientists from around the world are issuing calls to action—such as protecting the food supply and reducing exposure to chemical cocktails in the environment—to protect the fertility and reproductive health of future generations.

9

IMPERILING THE PLANET:
It's Not Just about Humans

Soiling Our Nests

In the North Pacific Ocean lies an enormous trash vortex, a convergence of more than eighty-seven thousand tons of floating debris, including plastic particles, chemical sludge, and other fragments of litter. This mass of detritus has come to be known as the Great Pacific Garbage Patch. Rather than being a discrete mass like an island, this swirling maelstrom of refuse is more like a diffuse galaxy of garbage that has grown to be about twice the size of Texas. It poses a danger to wildlife since the debris often ends up in creatures' stomachs or wrapped around their necks. Of the 1.5 million albatross that inhabit Midway Atoll near the Garbage Patch, the vast majority have plastic particles in their digestive systems, and approximately one-third of their chicks die. The floating debris absorbs organic pollutants in the seawater, then fish and other marine life consume these toxin-containing pieces of plastic. When humans eat these fish, we ingest microparticles of these toxic chemicals—another harmful trickle-down effect in our ecosystem.

This is hardly an isolated occurrence. There's also a tide of plastic waste along what was once the idyllic coastline of the tiny Honduran island Roatán, as well as a nearby series of "trash islands," composed of garbage, particularly Styrofoam and plastic, along with seaweed. In 2017, a floating mass of tiny plastic pieces larger than Mexico was discovered in the South Pacific. Meanwhile, in the Atlantic Ocean, "extreme" concentrations of microplastic pollution have been found in the Sargasso Sea. And in 2019, researchers spotted a floating mass of plastic waste, dozens of miles long, between the islands of Corsica and Elba in the Mediterranean Sea.

In each of these locations, sea creatures are literally swimming in trash-and-plastic-chemical soups of one kind or another. The United Nations Ocean Conference estimated that the oceans may contain more weight in *plastics* than *fish* by the year 2050. Intentionally or not, human beings are treating the planet's oceans like a garbage dump.

The debris-strewn oceans aren't the only casualties of our throw-away society, and these masses of detritus aren't just unsightly. They're also harmful to the environment, especially given that plastics, in particular, take thousands of years to decay. By some estimates, plastic is killing more than a hundred thousand sea turtles and birds per year, whether it's because these creatures ingest them or become entangled in them. Meanwhile, the chemicals from plastics contaminate fish and enter the food chain, which means they can be passed from one species to another and affect human health, too. As the Environmental Protection Agency notes, "Wildlife also can act as sentinels for human health: abnormalities or declines detected in wildlife populations can sound an early warning bell for people."

But it's not just about us because the health and vibrancy of other species matter—to them, and to the health and integrity of the planet in general. The difference is, other species haven't chosen to bring these chemicals into their lives and their habitats. Humans have done it for them, which means they've been innocent victims of humans' reckless and feckless behavior.

As you've read, even when specific chemicals are banned, they can persist in the environment for years, where they can harm other creatures. These persistent chemicals include heavy metals such as lead and mercury, as well as arsenic, PCBs, DDT, dioxin, and others, all of which are known or suspected endocrine-disrupting chemicals (EDCs). And as is the case for humans, other species are often simultaneously exposed to numerous EDCs, which creates the potential for harmful additive effects. But it's not just a 1 + 1 proposition; these effects can interact in ways that make the combination, or the whole effect, even worse than the sum of the separate parts.

After all, phthalates are in plastics, PVC pipes, home furnishings, and personal-care products. Phenols are in antiseptics, disinfectants, and medical products, among others, while perfluorooctanoic acid (PFOA) is in carpets, fabric protectors, stain repellents, and Teflon pots and pans. This continuous exposure may be why these nonpersistent chemicals can easily be measured in urine and have been found in a majority of people in Western populations. While other species don't "use" these products, they are exposed to them through by-products formed during chemical manufacturing and combustion, global transport of these chemicals through ocean and air currents, electronics recycling and garbage, and other processes.

As the use of some persistent organic pollutants has decreased, the use of nonpersistent compounds has increased. Yet, both classes still pose risks to reproductive organ development and can cause adverse neurological, endocrine, genetic, and systemic effects in humans and other species.

Body Burdens in Animals

Unfortunately, these ubiquitous environmental chemicals have taken a toll on the animal kingdom in many different ways. A recent study found that 88 percent of biopsies from bottlenose dolphins from the northern Adriatic Sea had PCB concentrations above the toxicity threshold for physiological effects in marine mammals, and 66 percent had concentrations above the threshold for reproductive impairment. Meanwhile, exposure to organochlorine pesticides, PCBs, and brominated flame retardants had an adverse impact on the reproductive function of Baltic gray seals, including a high incidence of uterine fibroids in females, and led to marked declines in their population. Male polar bears in East Greenland with high fat-tissue concentrations of persistent organic pollutants, including organo-chlorine pesticides and PCBs, were found to have reduced testosterone levels, unusually short penises, and smaller-than-normal testicles. There's even a disorder called imposex, which causes *female* sea snails to develop male sex organs such as a penis and vas deferens.* The cause: exposure to certain marine pollutants, particularly tributyltin (TBT), a highly toxic chemical that had been used extensively to prevent the growth of marine organisms on the hulls of large ships.

The point is: The effects of the chemicals we have unleashed into the world are vast and far-reaching, endangering the reproductive health of numerous species and possibly their very survival.

Case in point: in a series of studies, University of California, Berkeley, developmental endocrinologist Tyrone Hayes, PhD, investigated the effects of atrazine, an herbicide that's used primarily on corn, soybeans, and other crops in the Midwest and around the world, on the sexual development of wild leopard frogs. He found that exposure to atrazine had a feminizing effect on male frogs, leading to gonad abnormalities such as the presence of eggs in their testicles and testosterone levels that are lower than in normal female frogs. Toads have been found to have similarly dysfunctional reproductive responses to various EDCs. Given these reproductive abnormalities, is it any wonder that frogs and toads are undergoing a precipitous population decline throughout the world?

One of the most dramatic and widely reported examples of this kind of chemical impact on wildlife came from central Florida. For many years, Lake Apopka, one of the largest freshwater lakes in Florida at 12,500 hectares, was among the most contaminated lakes in the state. This was due to pesticide use in agricultural activities around the lake, a nearby sewage treatment facility, and a 1980 major pesticide spill, of a mixture of dicofol, DDT and its metabolites, and sulfuric acid, from the

---

*   Remember, the vas deferens is the duct that conveys sperm from the testicle to the urethra.

former Tower Chemical Company, which was adjacent to the lake. These pesticides can act as estrogens, binding to and activating estrogen receptors and inducing cellular growth that's estrogen-dependent.

In the 1990s, University of Florida wildlife biologist Lou Guillette Jr., PhD, and his colleagues compared the reproductive development in juvenile alligators from Lake Apopka and those from a (clean) control lake, Lake Woodruff, in central Florida. Going out on the lakes at night in teams in airboats, the researchers would catch baby gators and take various body or body-fluid measurements; or they would collect eggs from nests during the day. They found that at six months of age, baby female alligators from Lake Apopka had blood estrogen levels that were nearly twice that of the female alligators from uncontaminated Lake Woodruff—and it was clearly not because the female alligators were taking estrogen of their own volition. The Apopka female alligators also had altered reproductive tract development, including more abnormalities in their eggs and ovarian follicles (similar to what happens with PCOS in human females).

It wasn't just the females who were having reproductive troubles. The young male alligators from Lake Apopka had their own set of problems, particularly abnormally small penises and poorly organized seminiferous tubules (where sperm cells germinate and mature before being transported) in the testicles. What's more, the Lake Apopka male alligators had significantly lower testosterone concentrations—levels that were three times lower than those of the male alligators from Lake Woodruff and comparable to those of the *female* alligators from Lake Woodruff. Not surprisingly, these abnormalities had the potential to significantly thwart normal sexual maturation and the alligators' prospects of successfully reproducing.* Even in the wild on Lake Apopka, the hatching success rate of alligators was only 5 percent, compared to the 85 percent success rate it should be in a less contaminated lake.

These discoveries were disturbing in their own right, but they also pro-

---

* The low testosterone alone may have thwarted the male alligators' interest in sex.

vided telling insights about the risks of human exposures. Alligators have a similar life span to that of humans and can also reproduce for decades. So these researchers were able to learn about the effects of pollutants on reproduction that could be relevant to humans, even though we don't literally swim in a toxic soup.

But such adverse effects from exposure to chemicals are hardly limited to creatures residing in bodies of water. On land, Florida panthers that were exposed to high concentrations of DDE, mercury, and PCBs were found to have lower sperm density, motility, and semen volume, and higher numbers of abnormally shaped sperm compared to other panther populations. In Canada, researchers obtained 161 mink carcasses from commercial trappers in the provinces of British Columbia and Ontario between 1998 and 2006 so that they could examine the effects of EDCs, including organochlorine pesticides, PCBs, and polybrominated diphenyl ethers (PBDEs), on the males' reproductive development. The researchers found a significant relationship between DDE levels in the livers of adult mink and their penis length and size, most likely because DDE is antiandrogenic. Furry creatures were just as likely to suffer the reproductive fallout from these chemicals as those with scales.

The Dramatic Fall of Insects and Birds

In recent years, we've been hearing dire warnings about what's being called an insect apocalypse. A 2017 study from Germany found that the country's nature reserves had experienced a 75 percent decline in flying insects over the previous twenty-seven years. In coastal areas of California, the population of Western monarch butterflies plunged by 86 percent from 2017 to 2018. In Puerto Rico, the abundance of arthropods—including insects that have exoskeletons (such as beetles), as well as spiders and centipedes—has been declining at a disturbing rate and so have the populations of the lizards, frogs, and birds that eat them.

Whether or not you appreciate insects or fear them, the simple reality is, *we cannot survive without insects.* As the American biologist, naturalist, and author E. O. Wilson famously noted, "If all mankind were to disappear, the world would regenerate back to the rich state of equilibrium that existed ten thousand years ago. If insects were to vanish, the environment would collapse into chaos." Insects pollinate plants and trees and provide food for birds and other animals. Cows couldn't survive without grass, and grass wouldn't exist if beneficial insects didn't provide a natural form of pest control for those insects that damage grass, and help with the breakdown of organic matter so nutrients can be returned to the soil. Some species of fish wouldn't exist if they didn't have insects to eat. And chickens depend on insect-pollinated plants for the seeds and nuts they feed on. Insects are an integral part of the circle of life.

Among the suspected reasons for the demise of various insect populations: climate change and the widespread use of herbicides and pesticides. The global decline in the populations and diversity of insects has the potential for significant ripple effects on "food webs," the interconnected food chains within an ecological community, and hence the survival of various ecosystems.

Since 1970, North America has lost nearly 3 *billion* birds, a 29 percent reduction, across hundreds of species from warblers and finches to swallows and sparrows, according to a 2019 study. This is a crisis because birds also are a critical part of both the natural food chain and the planet's ecological integrity. While the degradation of high-quality habitats is the single greatest cause of the bird declines, according to Michael Parr, president of the American Bird Conservancy, pesticides are a contributing factor. Since DDT was banned or phased out, another pernicious generation of pesticides called neonicotinoids has been introduced.* As Parr wrote in a September 2019 opinion piece in the *Washington Post*, "Neonics are used to inoculate plants against insects.... They remove both harmful and beneficial insects. If you use a billion pounds of insect

---

*   This is another example of a regrettable substitution.

poison annually—as we do on the American landscape—you are going to wind up with fewer and fewer insects. Then fewer birds."

This is already happening off the northwest coast of Iceland, where things are uncharacteristically quiet these days. In recent years, colonies of puffins, kittiwakes, terns, and other bird species have been dying off or disappearing, and so have their chipper choruses. The numbers of (penguin-like) thick-billed murres dropped 7 percent per year between 2005 and 2008, while the populations of common murres and Atlantic puffins decreased considerably between 1999 and 2005, according to a 2016 report from the UN. It's not just that they're dying off at a faster rate; they're not reproducing at the rate they once did, either.

A major reason for this unfortunate demise: our high-carbon life-styles are turning up the oceans' temperatures, changing their chemistry, pollution loads, and food webs, and jeopardizing the health of various forms of marine life. Levels of "forever chemicals" such as PCBs and brominated flame retardants are taking a toll on these populations as well. The plight of these seabirds is sounding a warning bell throughout the world that more patterns like this are likely to be seen in the future. Once again, we, humans, created these fatal and fertility-altering effects.

Hijacking the Mating Game

Meanwhile, some environmental contaminants have been found to alter the mating and reproductive behavior of certain species. We've seen alterations in courtship and pairing behavior in white ibises that were exposed to methylmercury, the most toxic form of mercury, in Florida. One study found a significant increase in homosexuality in male ibises that were exposed to methylmercury, a result the researchers attribute to a demasculinizing pattern of estrogen and testosterone expression in the males; sexual behavior in birds (as in humans) is strongly influenced by circulating levels of steroid hormones including testosterone.

We are also seeing changes in reproductive behavior among female freshwater fish that are exposed to androgenic endocrine-disrupting chemicals; simply put, these female fish spend less time associating with their male counterparts. In other instances, both sexes can have their sexual behavior hijacked by environmental exposure to EDC chemicals. A case in point: trenbolone acetate is an anabolic steroid (similar in action to testosterone) that is widely used in some parts of the world to increase muscle mass in livestock; it used to be popular in the bodybuilding community, but it has been banned from human use.

Unfortunately, several metabolites of trenbolone acetate have been found in aquatic systems that are near animal feedlots. Researchers have found that fish that are exposed to even low concentrations of this androgenic chemical can experience disruptions to their reproductive development and function; in particular, female fish become masculinized during their early development, and adult females can experience detrimental effects on their fertility. In another wrinkle, a study from Australia found that short-term exposure to trenbolone altered the courtship and sexual behavior of male guppies, as well as the female guppies' receptivity to the males' sexual advances.

Other Dangers in the Water

In the Western world, people expect their drinking water to be safe, which is why the 2016 lead-contamination water crisis in Flint, Michigan, and the more recent one in Newark, New Jersey, elicited such strong public and political outrage. But an often overlooked reality is that, in addition to the possible presence of toxic metals, pharmaceutical drugs, including oral contraceptives and other hormones, may be lurking in our water supply, as well as in waterways that are home to fish and other creatures.*

---

* It's widely recognized that use of pharmaceutical drugs has increased dramatically in the United States and other countries in recent years. Even when adjusted for inflation, spending on retail prescription drugs increased from $90 per person in 1960 to $1,025 in 2017 in the United States alone.

Regrettably, the chemicals in these drugs end up in waterways after being excreted from the human body or when unused medications are flushed down the toilet. These drugs can also enter our waterways through manufacturing waste, animal excretion, runoff from animal-feeding operations, or leaching from municipal landfills, according to a report from the Natural Resources Defense Council (NRDC). What's more, medications excreted in human urine, feces, and bathwater can migrate from sewers into oceans, rivers, lakes, and streams, where they can harm various forms of wildlife.

As a result, it's hardly surprising that drug-polluted waterways are now home to a variety of intersex fish—namely, males that produce eggs. Or, that fish and shrimp living in water that contains traces of anti-depressants show alterations in their normal behavior such as staying at the water's surface or swimming toward light, either of which can make them vulnerable to predators. Meanwhile, fathead minnows that have been exposed to antidepressants and anticonvulsants in water have exhibited neurological changes, some of which resemble autism-like disorders.

Facing the Messes We've Made

This should give you a pretty clear picture of what's going on—and what's going wrong—with other species throughout the world. When the chemicals that we humans have created seep into the environment, they can take a toll on the health, development, behavior, and even survival of other creatures. The bottom line: we're essentially dosing the entire planet when we take these drugs or dispose of them improperly. Other creatures didn't sign up for this.

Making matters worse, the chemicals that are altering *our* reproductive development and function, as well as that of the alligators, frogs, and other species, are largely coming from industries that are damaging our climate as well. As a panel of one hundred endocrine-disruption and climate-change scientists wrote in a 2016 commentary in *Le Monde*, *"Many of the actions*

*needed to reduce the burden of endocrine disruptors will also help in the fight
against climate change.* Most man-made chemicals are derived from fossil fuel by-products manufactured by the petrochemical industry.... *These
chemicals compromise male reproductive health and contribute to cancer risks.**

Already, there is concern that exposure to EDCs may hinder the ability of other species to adapt to environmental changes that are driven
by climate change, given that EDCs alter hormonal programming and
function. As Norwegian scientist Bjørn Munro Jenssen, who studies how
environmental pollutants affect animals, wrote, "When taking into consid-
eration the long-range transport of EDCs into the Arctic ecosystem, the
combination of EDCs and climate change may be a worst case scenario
for Arctic mammals and sea-birds."

In the past, the presence of chemicals in the environment was regulated primarily on the basis of what causes cancer, but the levels that
threaten reproductive health are usually lower. This means that regulating
chemicals on the basis of cancer risk can miss significant reproductive
risks. For example, when the US EPA analyzed fish tissue from 540 river
sites across the country, the screening value for noncancer endpoints,
including reproduction, was four times higher than that for cancer. The
concentration of twenty-one PCBs was found to exceed the level considered to pose an increased risk of cancer in humans in 48 percent of the
samples; this likely means that the thresholds for reproductive damage
have already been met. Findings like this suggest that it's time for a new
set of regulatory standards, ones that will protect reproductive development and function for all living creatures.

Ultimately, whether it's through our lifestyles or the chemical contaminants we have developed and unleashed, we're imperiling the world
in which we live. Where the effects will stop is unknowable—unless we
take crucial steps to reverse the exposures to chemicals in our midst and

---

*   As the scientists noted, by reducing reliance on fossil fuels and shifting to
alternative forms of energy, greenhouse gas emissions could be reduced, which would
help with the climate crisis; this would also reduce the production of chemical products
that can harm the reproductive health of men, women, children, and other species.

the burdens these chemicals are having on other living creatures. While it's true that environmentally induced reproductive disorders in other species are important sentinels for men's and women's reproductive health, the sexual development and functionality of other species matter in their own right. This isn't an us-or-them proposition. We're all surrounded by the same toxic stew. There's simply no place on the planet that's safe from these chemicals.

We created these problems, albeit unwittingly, so it's up to *us* to come up with the solutions, as you'll see in subsequent chapters. Although limited so far, government actions to ban or restrict the use of potentially harmful chemicals, in order to reduce exposure to them, have already contributed to decreases in the frequency of certain disorders in wildlife, as the 2012 WHO report acknowledged. For example, after a decline in the environmental concentrations of PCBs and organochlorine pesticides, the populations of Baltic Sea seals, which had previously had a high incidence of fibroids associated with exposure to these chemicals, has been rebounding. Since TBT was banned in 2008 from use in marine antifouling paints, the populations of marine gastropods have been recovering throughout the world; and in 2017, no signs of ambiguous genitalia were found among sea snails in any of the monitoring stations along the Norwegian coastline. These are important examples of how cleaning up the environment can remove threats to reproductive development.

Unlike other species, we, as human beings, have the choice and the ability to take steps to reverse these harmful influences. Altering this downward trajectory is likely to require drastic changes in our collective lifestyle and our regulatory processes for chemicals, pharmaceuticals, and consumer products. The challenge may seem akin to turning around the *Titanic*. But it can be done and it's worth the effort, because the health, vitality, and longevity of the human race, other species, and the planet depend on it.

10

IMMINENT SOCIAL INSECURITIES:
Demographic Deviations and the Unraveling of Cultural Institutions

Replacement Values

When people hear about the precipitous decline in sperm counts that has taken place in Western countries, some shrug it off and say, "Well, the world is overpopulated; fewer kids is a good thing." But that's not necessarily true. Western cultures are experiencing a "demographic shift"—their populations are aging, and with birth rates dropping, these countries are not replacing their populations. This is even more true during the age of COVID-19. It would require couples to have an average of 2.1 babies to sustain a country's population through new births alone. But in most Western and in some Eastern countries, that benchmark isn't being achieved.

In the United States, for example, the fertility rate, which is defined as the average number of children born per woman, was 1.8 in 2017, a 50 percent drop from 1960, according to data from the World Bank. In 2018, the United States had the lowest number of births in thirty-two years! In Canada, the fertility rate dropped from 3.8 in 1960 to 1.5 in 2017. In Italy and Spain, the fertility rate is now down to 1.3. In Hong Kong it has plummeted from 5.0 in 1960 to 1.1 in 2017, while in South

Korea it has dropped from 6.1 in 1960 to 1.1 in 2017. And the number of babies born in 2019 in China fell to its lowest point since 1961, triggering what's being called "a looming demographic crisis." A major analysis from the Global Burden of Disease Study corroborates these worldwide findings. Using fertility data from 195 countries and territories, after accounting for mortality and migration rates, the researchers found that the total fertility rate decreased in all the countries included in the study and declined globally by 49 percent between 1950 and 2017. (If you're suffering from stat overload, sorry about that, but I want you to have a sense of the scope and magnitude of these shifts.)

This is a sea change. For many years, the world's population seemed to be rising at a steady clip. If the world's average fertility rate in 1970 had remained consistent and still held true today, the global population would be 14 billion, or nearly double what it is currently. But things didn't play out that way. While the decline in sperm counts in Western countries has undoubtedly played a role in this decreasing fertility rate, other factors are influencing these shifts, too. In the United States and many other countries, men and women are waiting longer to get married, and they're having their first child at an older age, which leads to smaller families. Once people start having fewer children, they're unlikely to stop because they may discover that having fewer offspring is easier to manage and more affordable.

A leading cause of this downward fertility trend, according to a 2018 report on global fertility rates, reflects the increase in women's choices, which have grown exponentially in some parts of the world. In particular, increases in women's education levels and women's reproductive rights, which include the availability of contraceptive methods around the globe, are driving the declining birth rate. The correlation between a young woman's educational opportunities and the number of children she's likely to have is clear throughout the world, but it is particularly noteworthy in countries where historically girls didn't have the same educational opportunities as boys. A 2015 study by researchers from the Harvard School of Public Health examined the effects of schooling on teenage fertility in Ethiopia, based on education reform policies that were introduced in 1994. The researchers found that each additional year of school led to a 6 percent reduction in the probability of teenage marriage and teenage childbearing.

Similar relationships have been found between increases in female education and lower rates of early childbearing in Indonesia, as well as in Nigeria, Ghana, Kenya, and other countries in sub-Saharan Africa where, historically, the gender gaps in secondary school enrollment between boys and girls have been sizable. What's more, between 1950 and 2016, dramatic declines in the birth rate that occurred in the Republic of Korea and Singapore coincided with heavy investments in education for girls, efforts to increase women's participation in the workforce, and high rates of urbanization.

Indeed, urbanization has been acknowledged as a significant factor in the fertility decline of recent decades. Between 2011 and 2015, women living in rural areas in the United States were 32 percent more likely to have had three or more births than women in urban areas. This may be partly because in rural areas children are often viewed as valuable commodities, as part of the (free) labor force that can work in the fields, feed the cows or horses, collect eggs, or handle other essential chores. In cities, by contrast, kids, beloved as they are, become more of a financial burden than an asset—another body to feed, clothe, educate, and rear, all of which is generally more expensive in a city or suburban environment than in rural areas. Given that between 2000 and 2016 the share of people living in urban areas in the United States has remained steady, while the proportion has increased in suburban and small metropolitan areas and decreased in rural areas, it's not surprising that the nation's fertility rate has been declining.

Global Population Ups and Downs

Despite the downturn in birth rates in the Western world, a large swath of the world still has fertility rates above replacement levels. In Chad, it's

5.8. In the Congo and Mali, it's 6.0. And in Somalia, it's 6.2. So while the fertility rate is declining in some parts of the world, it's still high in other regions, particularly certain African countries, which is why the world's population is currently increasing. Nevertheless, the planet's population growth isn't likely to continue the way demographers once predicted.

The United Nations Population Division has developed various scenarios, based on statistical models, to project the growth trajectories of the world population. Of particular interest are three scenarios called the high, medium, and low variants (or growth forecasts). The medium variant, which many demographers consider to be most likely to play out over the rest of the century, is the middle-of-the-road scenario. In 2019, the UN's medium variant estimated a world population in 2100 of approximately 11 billion. By contrast, the high variant projection is based on a higher forecasted birth rate than the medium variant, while the low variant prediction reflects birth rates that are lower. Under the high variant scenario the world's population would be 15.5 billion in 2100, nearly double what it is today. The low variant predicts a world-wide fertility rise and fall, in which the global population will peak at 8.5 billion in 2050 and then (surprisingly!) decline to around 7 billion at the end of the century.

While the medium variant scenario is widely quoted, some demographers and population experts disagree with this projection. Jørgen Randers, PhD, a Norwegian academic who coauthored the 1972 book *The Limits to Growth*, once warned of a potential global catastrophe caused by overpopulation. He has since changed his mind. In a 2014 TEDx Talk, he stated, "The world population will never reach nine billion people. It will peak at eight billion in 2040 and then decline." Randers believes the primary driver of this decline will be that the world's women will choose to have fewer children than in the past.

Other experts echo his beliefs. For example, a 2013 Deutsche Bank report suggested that the planet's population will peak at 8.7 billion in 2055 and then fall to 8 billion by 2100. Demographer Wolfgang Lutz, PhD, founding director of the Wittgenstein Centre for Demography

and Global Human Capital in Vienna, Austria, believes that populations that are experiencing low fertility rates are caught in some kind of
"low-fertility trap." The gist of his hypothesis is that "once fertility has
fallen below certain levels and stayed there for a certain time, it might
be very difficult, if not impossible, to reverse such a regime change." This
hypothesis is based on three independent elements. As a society experiences a drop in fertility rate below replacement level, there will be fewer
women of childbearing age, which means the number of subsequent
births will decline; new generations embrace a smaller ideal family size,
based in part on the lower fertility they see with previous cohorts, which
creates sociological reinforcement; and third, assuming that the aspira-
tions of young adults are on an upward trajectory, their expected income
isn't likely to parallel this rise, which makes the prospect of having fewer
children feel more realistic. In Lutz's view, these three factors will contribute toward "a downward spiral" in the number of future births.

In Some Ways Age Is More Than Just a Number

The demographic picture for the United States and the rest of the world
today looks quite different from the way it has in recent decades—and
this trend is expected to continue. "Growth from 1950 to 2010 was
rapid—the global population nearly tripled, and the U.S. population
doubled," as a 2014 Pew Research Center report noted. "However, population growth from 2010 to 2050 is projected to be significantly slower
and is expected to tilt strongly to the oldest age groups, both globally and
in the U.S."

We are already seeing sizable shifts in this direction: In 1960, 5 percent of the world's population was sixty-five and older; in 2018, that
proportion rose to 9 percent, according to the World Bank. Similarly, 9
percent of the US population was sixty-five and older in 1960; in 2018,
the proportion rose to 16 percent. And in the twenty-eight countries that
make up the European Union, 10 percent of the population was over

sixty-five in 1960, while in 2018 that figure rose to 20 percent. Everywhere in the world, the over-sixty-five population has nearly doubled from what it was in 1960.

As the birth rate declines and life expectancy increases, the population of elderly people continues to grow throughout the world. Life expectancy in the United States is now seventy-nine years, up from seventy in 1960. In Japan and Switzerland, it's now eighty-four, compared to sixty-eight and seventy-one, respectively, in 1960. Admittedly, these increases in life expectancy are one of the great achievements of the twentieth century. But the declining birth rate is not. That shift is the opposite of what was happening a century ago when sperm counts and fertility rates were high and life spans were considerably shorter.

This is where the aforementioned "demographic time bomb" comes in—population experts and scientists fear that future generations will struggle to meet the needs of an ever-increasing number of older adults and retired workers and their pension/social security obligations. Regarding countries where the fertility rate has dropped, particularly in North America, Asia, and Europe, the United Nations Population Fund's report *State of World Population 2018* noted, "With larger groups of older people and a shrinking labour force, these countries face potentially weaker economies in the near term."

In most developed areas of the world, the proportion of older adults already exceeds that of children, and by 2050, one in six people in the world will be over age sixty-five, an increase from one in eleven in 2019. There will be far fewer people of working age to support those who are over sixty-five. As the population ages, the ratio of older adults to working-age adults (defined as those twenty to sixty-four) is projected to rise. In the United States, for example, in 2020, there are about 3.5 working-age adults for every adult of retirement age; by 2060, that proportion is expected to shrink to 2.5. Since economically active people pay considerably more in income tax and other taxes, while economically inactive people, such as children and older adults, tend to be bigger recipients of government spending in public education, health care, and pensions, an increase in the dependency ratio would cause fiscal problems for the government of that country.

The potential impacts of these shifts "are beyond huge," says Darrell Bricker, PhD, a demographic commentator and coauthor of *Empty Planet: The Shock of Global Population Decline.* "There are questions about how to support an aging population and a need to rethink all aspects of how public money is spent on pensions, health care, city infrastructure, schools, the military. Those are young people's games. What happens when there aren't enough young people? Who's going to pay for retirement? When you have a consumption-based economy, what happens when your population is old and the wealth resides with the older generation?"

These shifts have many potential consequences for societies. These include "reductions in economic growth, decreasing tax revenue, greater use of social security with fewer contributors, and increasing health-care and other demands prompted by an ageing population," according to the 2017 Global Burden of Disease Study. In the United States, a doubling in the number of adults over sixty-five, which is expected by 2060, could lead to more than a 50 percent increase in the number of older adults requiring nursing home care by 2030, according to the Population Reference Bureau. How we manage these changes will have significant implications not only on the economy, but also on our culture, politics, and nearly every aspect of society.

In the United States, these changes could lead to an "epic" crisis for Medicare and Social Security, warns Daniel Perrin, a nationally known policy leader and lobbyist on health care, public debt, and senior issues. After all, both programs are financed through taxes that are tied to workers' earnings; a decline in the working-age population could drain the financial reservoirs of these resources. Yet, Perrin says, many people aren't aware of these demographic shifts, and "those that do know about this have a hard time wrapping their minds around it. They have a hard time squaring this with human history." As a result, policy makers in the United States are unprepared for these population shifts and the economic and social-support challenges that are sneaking in along with them. For the year 2091, the Social Security Administration predicts that expenditures will exceed income by at least 4.48 percent and possibly increase to 5.97 percent if fertility rates stay low. It doesn't take a math whiz to see how that's problematic for the sustainability of this social-support institution.

Research suggests that a country's peak potential for economic growth occurs when the proportion of the population that is of working age (fifteen to sixty-four) is larger than the proportion that is of nonworking age—such a country is said to reap a demographic dividend. This is true throughout the world, but changes are afoot on this front, too. Since the 1960s, the proportion of the population that's of working age increased in high-income countries, crossed the significant 65 percent threshold in the late 1970s, then remained relatively steady for the next two decades. Things started to change in 2005, as the proportion of working-age people began to decline in these countries, and as of 2017, in twelve of thirty-four high-income countries throughout the world, the proportion of the population of working age is less than 65 percent. That's problematic on many levels.

Shifts such as these can have profound implications for the economic vibrancy, as well as the cultural and social conditions, of a particular region. In these countries, changes in the proportion of working-age adults to older adults could have such a significant effect on economic productivity that it will likely lead to later retirement, well after age sixty-five, which is already occurring in the United States, Australia, and Japan. These changes mean that by the time *you're* over sixty-five, you may not be able to draw Social Security payments or Medicare or have access to the health care you need if there aren't enough people to provide it.

Notably, Japan's proportion of the population that's of working age has dropped to less than 60 percent. In Japan, the sixty-five-plus proportion of the population was 6 percent of the total in 1960 and surged to a whopping 27 percent in 2018. These days, there aren't enough health-care workers to care for the elderly population (and restrictive immigration laws aren't helping). Meanwhile, the birth rate is down to 1.4, sperm counts are low, and fewer males are being born compared to females, as often happens in response to environmental stressors.

At the same time, more women of childbearing age are putting their careers first and postponing or rejecting marriage and motherhood. The Japanese culture places such a premium on professional success and long work hours that many people of reproductive age reportedly aren't even interested in having sex, according to multiple sources. This has supposedly given rise to "celibacy syndrome" (*sekkusu shinai shokogun*), which has been described as a decline in sexual interest and activity or even romantic relationships among young adults in Japan.

The reasons for this sexual slump aren't well understood. As an article in the *Independent* noted in 2017, "The fertility crisis has left politicians [in Japan] scratching their heads as to why youngsters are not having more sex." Naturally, there are theories, ranging from the enduring social values of modesty and purity in Japan (which make the prospect of casual sex difficult to navigate), to young Japanese men's and women's changing life desires—for example, being more dedicated to careers, not wanting to have traditional relationships, and showing an increased interest in online pornography. Whether hormonal factors or dietary influences play a role is a matter of conjecture—but some evidence suggests that testosterone is lower among Asian people and that greater consumption of soy foods, which are rich in estrogenic compounds, may have a libido-compromising effect in men. It may be that, in Japan, a perfect storm of physiological, cultural, dietary, and environmental influences is leading to a loss of that loving feeling (not just lower sexual frequency but also lower sexual satisfaction).

Interestingly, this conscious or unconscious uncoupling—along with the so-called epidemic of loneliness that's been identified in the country—has spawned some new social inventions to help people feel less alone. In Japan, anyone who wants to have a child, without actually having one, can buy a toy-size robot companion that has the mental acuity of a fifth grader. For $3,000 or more, men can purchase lifelike anatomical female (sex) dolls for companionship. It's not unusual to see men taking these dolls for walks in wheelchairs in public. The Japanese artist Tsukimi Ayano has been crafting mannequins and positioning them throughout the tiny village of Nagoro, in southern Japan, in an effort to make the place feel more populated as people move away or die. Recently an industry has sprung up that allows lonely people to "rent" family members—actors who play the roles of spouses, parents, children, or grandchildren—for temporary companionship. One of the hazards of the occupation: client dependency. Sometimes clients just don't want to say goodbye to these rented relatives.

A salon owner in San Francisco, Shiori, forty-three, was raised in Japan and came to the United States in 2001. Married with two children, she and her family travel to Japan every few years to visit relatives, including Shiori's younger sister, who is single and doesn't want to have children. While visiting Japan in August 2019, Shiori was struck by the sense that "people are lonely. Country schools have been reduced to one-room schoolhouses because there are so few children. Instead of dating, young adults prefer to relax by going to a manga café or an internet café."

The population of Japan has been shrinking steadily since the 1970s. By 2065, it is expected to drop to about 88 million, compared to 126.5 million in 2018. With fewer babies being born and a growing elderly population, Japan is facing the prospect of an unparalleled demographic crisis that could have significant ripple effects socially, economically, and politically. To try to avert this looming crisis, some local governments in Japan have been offering cash incentives to encourage young women to get busy having babies. While some evidence suggests that this approach has spurred a slight uptick in the fertility rate in certain areas, whether it will last remains to be seen.

The situation in Singapore is equally disturbing. The most recent figures put the total fertility rate at 1.1. In 2018, the personal lives of Singapore citizens were examined in detail in the country's parliament, as members wrung their hands over the country's low birth rate and wondered why government schemes to encourage parenthood hadn't produced more results. A minister said Singapore's total fertility rate had fallen below replacement levels for some forty years, noting that these same trends have played out in developed East Asian societies such as Japan and South Korea. The parliament recognized that financial and legislative measures alone aren't enough to turn things around.

When a popular online publication solicited ideas from readers for ways to improve Singapore's birth rate, all the suggestions related to improving social support, financial incentives, access to child care, and free fertility checks—and encouraging Singaporeans to have more sex, which, surveys suggest, they aren't doing on their own. One thirty-two-year-old man suggested, "The parliament should start a campaign to make it fashionable to have sex." Another suggestion: "The best role for women is at home," which suggests that one backlash to low fertility rates, at least in Singapore, is to try to keep women out of the workplace and have them stay home to raise children instead.

What's happening in Singapore and Japan provides a cautionary glimpse into the future for the United States and other countries that have declining fertility rates. So far, Japan and Singapore have been unable to turn their birth rates and population declines around. In the United States, we're on the same trajectory, and we may end up facing similar challenges.

Which Sex Is Outnumbered Now?

Around the world, the ratio of men to women is changing, too. Historically, 105 males have been born for every 100 females, which means that 51.5 percent of births have been male. This is called the secondary sex ratio,* and this is what the World Health Organization expects the ratio of males to females to be at birth—it's considered the natural equilibrium, in other words. But this ratio isn't stable; it's influenced by biological, environmental, social, and economic factors.

---

* The *primary* sex ratio is the ratio of males to females at the time of conception, whereas the *secondary* sex ratio is the ratio at the time of birth.

Why this matters: The sex ratio can change in both human and wildlife populations in response to environmental factors and personal stressors. A shift in the sex ratio, which is usually in the direction of fewer male births, can be a sensitive indicator of sudden or pervasive environmental dangers. Surprisingly, a man's exposure to these dangers is more likely to lower the chances that his child will be a son than his female partner's exposure is.

As you saw in earlier chapters, while they're in the womb, males appear to be more sensitive to prenatal exposure to toxic chemicals, as well as to catastrophic events in the external world. Research has found that mothers who had the highest exposure to polychlorinated biphenyls (PCBs) from consuming contaminated fish from the Great Lakes were less likely to have a male child. And studies in Canada, Taiwan, and Italy have produced similar findings stemming from exposure to environmental toxins. (Remember, despite being banned in 1979, PCBs and other persistent organic pollutants, or POPs, continue to linger in our air, water, and soil; they're "forever chemicals" with the potential to do endless harm.)

Meanwhile, the 1995 Kobe earthquake in Japan, the 9/11 attack in New York, economic downturns, and war have all been shown to slightly lower the ratio of boys to girls that are born. In the case of the Kobe earthquake, some researchers suggest that "the sex ratio changes may be due to acute stress and a reduced sperm motility." (Fortunately, the effect on sperm motility is usually temporary and is typically restored to where it was within two to nine months.) Climate change also appears to be skewing the sex ratio: One study found that recent temperature changes—especially very hot summers and very cold winters—in Japan correspond to a lower ratio of male to female newborn infants, partly because of a dramatic increase in the proportion of male stillbirths. In particular, nine months after a very hot summer in 2010 and nine months after an especially cold winter in January 2011, more females were born than males.

It's not only external environmental factors that can affect a male baby's chances of surviving in utero. An expectant mother's stress level can also play a role. A study from Denmark found that among 8,719 pregnant women, those who experienced high or moderate levels of psychological distress in early pregnancy were less likely to give birth to baby boys. The mothers with the highest levels of psychological stress, based on their responses to a commonly used health questionnaire, had boys 47 percent of the time, whereas unstressed mothers gave birth to boys 52 percent of the time. This discrepancy may not seem like a big deal, but it means the difference between a sex ratio of .85 and 1.07— a considerable gap. The researchers concluded that stress during pregnancy is a likely culprit in the decreasing sex ratio in many countries.

While the biological mechanisms behind these effects aren't clear, some researchers suspect that after the twentieth week of pregnancy male fetuses may be more sensitive than female fetuses to a mother's corticosteroids, the hormones that are produced by the adrenal glands at higher levels in response to stress. This "elevated stress reactivity" could jeopardize the viability of males while they're in the womb. Regardless of the precise mechanisms behind these influences, given that males are especially threatened by environmental chemicals, climate change, and a mother's psychological stress, they will continue to face dangers in utero unless the world as we know it changes dramatically.

The Future Fallout Potential

All of these societal shifts should make us wonder, Who's going to run the show in the future if not enough children are being born to support the world that we've built? Who will take care of our older adults? What does this mean for the fate of the human race?

Whether it's because fewer males are born or women outlive men, the ratio of women to men will continue to increase, as part of the demographic shift, and an older population will be composed largely of women. And if the decline in sperm levels really is occurring at a faster rate in Western countries than in developing countries, as the current data suggest, there will be socioeconomic shifts throughout the world.

The world's population is in flux on multiple levels, and this much uncertainty is unnerving for the future of social-support programs, economic stability, national and international planning decisions, and other factors that are fundamental to a country's ability to operate efficiently. These shifts can affect the functionality of individual countries, as well as population shifts on the global stage. In 1950, high-income regions in central and Eastern Europe and central Asia accounted for 35 percent of the world's population; in 2017, the populations in these countries constituted 20 percent of the world's population. Meanwhile, large population increases occurred in South Asia, sub-Saharan Africa, Latin America and the Caribbean, North Africa, and the Middle East, as the Global Burden of Disease Study found.

When these trends are considered along with declining sperm counts, there's even more cause for concern; not only are *men* becoming endangered, but the human race as a whole is, too. Even if the will is there to reproduce and increase the birth rate, the machinery isn't as functional as it used to be, for men or women. We have declining sperm counts, diminished ovarian reserves, increasing miscarriage rates, and other reproduction-related problems that can hamper success in the realm of baby-making.

Some scientists are now suggesting that the detrimental effects on human reproduction, and the underlying factors contributing to them, could threaten the *survival* of the human race. It seems hard to fathom, but an argument could be made that *Homo sapiens* already fit the standard for an endangered species, based on the US Fish and Wildlife Service's (FWS) requirements. Of five possible criteria for what makes a species endangered, only one needs to be met; the current state of affairs for humans meets at least three.

The first is that we are arguably experiencing "destruction, modification, or curtailment" of our habitat. Our habitat includes our air, food, and water, each of which is being contaminated by pesticides, plasticizers, perfluorooctanoic acids (PFOAs), and other toxins that threaten human health and longevity. Nearly 25 percent of deaths worldwide—

which adds up to 12.6 million deaths every year—are linked to environmental issues, according to the World Health Organization.

The second FWS criterion met is that we have "an inadequacy of existing regulatory mechanisms"—given that our regulatory process assumes that most chemicals used in products are safe until they are proven to cause harm to humans, and, also, given that the testing methods behind these regulations are archaic.

And the third FSW standard we're meeting is that there are other "man-made factors affecting" our continued existence—including sharp increases in global temperatures. Presumably, you're familiar with the list of problems arising from climate change. What you may not know: It's suspected that global warming also contributes to decreasing sperm counts. In one study of semen quality in four European cities, sperm counts were 40 percent lower in the summer than in the winter.

This much is clear: Already, many populations aren't replacing themselves, sex ratios are changing, and marriage rates are going down—which creates a potential recipe for social and economic discordance the likes of which we've never seen. As climate change and environmental pollution persist, the ratio of male to female babies that are born will likely decrease further, and the proportion of adults over age sixty-five will continue to overshadow the under-fifteen crowd. It's hard to know what the future will look like for societies around the world.

Part IV

What We Can Do about This

11

A PERSONAL PROTECTION PLAN:
Cleaning Up Our Harmful Habits

American entrepreneur and motivational speaker Jim Rohn famously advised, "Take care of your body. It's the only place you have to live." That's absolutely true, of course, and only *you* can give your body the care it needs, from both the inside and the outside. As you've seen, lifestyle practices can affect reproductive health and functionality for both men and women, for better or worse. Some of the negative effects are reversible; others aren't—and the worst offenders are sometimes different for men and women.

If women want to have a baby, they are often told, "Clean up your act," but it's probably even more important for men to do so. For example, if you're a man, it's wise to steer clear of the hot tub, steam room, or sauna after your gym workouts, especially if you're trying to conceive, since exposure to intense heat can take a toll on sperm count and quality.* This effect is often reversible if men start avoiding these hot environments.

In some instances, women also can restore some of their reproductive

---

* If you're a pregnant woman, you should stay out of these extremely hot environments because they can cause you to become overheated or dehydrated, which can be harmful to the developing fetus.

health and functionality that was taken away by harmful habits. But if a woman's unhealthy lifestyle habits have gone so far as to harm her eggs, the damage is done and cannot be reversed.

Given what you read in chapter 6, you might think you need to start leading a monk-like existence for the sake of your reproductive health and fertility. But there's no need to take clean living to such an extreme. If you lead a generally wholesome lifestyle, you'll help safeguard your fertility and reproductive health over time. The good news is: when it comes to lifestyle factors, an easy rule is that what's good for your heart, mind, and immune system is also beneficial for your reproductive capacities. Fortunately, the health-protective strategies that are widely recommended for your overall health will also help protect your reproductive health.

While it can be challenging to improve your eating and exercise habits, particularly when life is hectic, do your best to follow these guidelines, without letting perfect become the enemy of good. The goal is to eliminate the unhealthiest of lifestyle practices and to develop healthier habits in other areas. Here's how:

Steer clear of cigarette smoke. If you smoke, quit—it's that simple. Smoking cigarettes is toxic to a man's sperm, and the chemicals in cigarettes, including nicotine, cyanide, and carbon monoxide, are toxic to a woman's eggs and speed up the rate at which those eggs die off.* Even if *you* don't smoke, being around secondhand smoke (aka passive smoking) could affect your reproductive health; this is especially true for women. So if someone in your household smokes, urge that person to quit or at least ban smoking inside your home.

Strive to maintain a healthy weight. That means a body mass index (BMI) between 20 and 25. As you've read, being substantially overweight or underweight has a negative effect on sperm quality, with obesity (a BMI of 30 or higher) being even more detrimental given its associations

---

*   Keep in mind that the jury is still out regarding the long-term effects of marijuana on reproductive health and functionality.

with lower sperm count, concentration, and volume, decreased sperm motility, and a higher incidence of shape abnormalities. Similarly, being considerably overweight or underweight (having a BMI under 18.5) can wreak havoc with a woman's hormone levels—causing irregular menstrual cycles or problems with ovulation or implantation of a fetus—and increase the risk of miscarriage if she is able to get pregnant.

If you're overweight or obese, make an effort to slim down by reducing your food (calorie) intake and increasing your calorie expenditure through exercise. Taking these steps to shed excess weight can make a difference if you're trying to get pregnant. Study after study has found that when overweight or obese women who are seeking fertility treatment adhere to a reduced-calorie diet and regular aerobic exercise, their pregnancy prospects can improve (in one study by 59 percent). Similarly, if underweight women gain weight or cut back on excessive exercise, in some cases their menstrual cycles may normalize, which will enhance their reproductive health.

Upgrade your diet. There's a sign I've seen several times that notes: THE KEY TO EATING HEALTHY? AVOID ANY FOOD THAT HAS A TV COMMERCIAL. It's sound advice because foods that aren't usually advertised on TV, such as apples and broccoli, or that don't have ingredients lists are generally more nutritious—and hence better for your overall health—than are packaged foods. (There's the added benefit of avoiding chemicals that are inherent in the packaging, as you'll see in the next chapter.)

People often want to know if there's a fertility-enhancing diet. The answer is, not exactly, but there's one that's close. Women who consume a Mediterranean-style diet—which emphasizes fruits, vegetables, whole grains, legumes, nuts, seeds, potatoes, herbs, spices, fish, seafood, skinless poultry, and extra-virgin olive oil—have been found to have a 44 percent lower chance of having difficulty getting pregnant. Research from the Netherlands found that couples who followed a Mediterranean diet before undergoing IVF/ICSI treatment had a 40 percent higher likelihood of achieving pregnancy than couples who adhered to other dietary patterns. What's more, research suggests that adherence to this kind of healthy diet is associated with better sperm quality in men and better fertility in women. An added perk: it can also help with weight management and promoting overall health.

It doesn't take long for a diet upgrade to make a difference in sperm. A 2019 study from Sweden found that after young, healthy men started following a wholesome diet—with yogurt, whole-grain cereal, fruits, vegetables, nuts, eggs, and the like—their sperm motility increased after just *one week*. Meanwhile, a higher intake of monounsaturated fats—from olive oil, avocados, and certain nuts—has been found to be associated with higher sperm concentration and total sperm count.

A high intake of omega-3 fatty acids has also been linked with improved semen quality and reproductive hormone levels in men,[*] as well as a reduced risk of ovulatory problems and improved fertility in women. The potential hitch is that some sources of fish and seafood are high in mercury, which is a concern for the fetus's developing brain in utero. To avoid mercury in seafood, put king mackerel, marlin, orange roughy, shark, swordfish, and tilefish on your no-buy list; stick with wild salmon, sardines, mussels, rainbow trout, and Atlantic mackerel.[†]

Compelling research suggests that vitamin D may be emerging as a major player in reproductive health. It's been shown to improve male fertility potential mostly by having a positive effect on sperm motility. And it's been found to improve sexual function and satisfaction among women with problems in that arena. In addition, vitamin D *deficiency* has been found to be much higher among subfertile women, which is why optimization through diet and possibly supplements is recommended.

Keep moving. Besides helping you manage your weight and stay fit, regular aerobic exercise and strength-training workouts are beneficial for

---

[*]   There's even some preliminary evidence that regularly taking fish oil supplements can improve overall testicular function in young men.

[†]   Hot tip: To reduce your exposure to PCBs in fish, remove the skin and visible fat before cooking. Grill, broil, or bake the fish, and let the fat drip off during cooking.

your reproductive function. It's true whether you're a man or a woman. Physical activity is beneficial for the production and virility of sperm, as well as being healthy for the rest of a man's body. In the Rochester Young Men's Study, we found that healthy young men who engaged in regular moderate to vigorous physical activity and watched less TV had higher sperm counts and sperm concentrations than less active men did. The most startling finding: men who performed moderate to vigorous exercise for fifteen or more hours per week had a 73 percent higher sperm concentration than those who got the least amount of exercise. Admittedly, that's a *lot* of exercise—slightly more than two hours per day, which is prohibitive for many guys who have busy work schedules.

Fortunately, this isn't an all-or-nothing proposition, because other research suggests that men who get more than seven hours per week of moderate to vigorous physical activity have 43 percent higher sperm concentrations than those who exercise an hour or less per week. More recently, a study of potential sperm donors in China found that men with the highest levels of moderate to vigorous physical activity have significantly higher sperm motility.

More good news: Men who aren't currently in the exercise habit should take heart because it's not too late to start. Research has found that when men who were sedentary and obese took up exercising at a moderate intensity on a treadmill for thirty-five to fifty minutes, three times per week, their sperm count, motility, and morphology improved after sixteen weeks. That's a relatively short-term investment in fertility potential.

The bottom line: moderate exercise is a healthy source of physical stress, whereas excessive exercise tips the balance into overload territory. This is true for men, and the same golden mean—Aristotle's term—applies to women.* Regular physical activity has been found to improve

---

* In his discussion of virtues and moral behavior, Aristotle focused on the middle state or golden mean between the extremes of excess and deficiency. I would argue that the same notion applies to lifestyle factors such as exercise, diet, and stress—an inverted U-shaped curve characterizes the optimal zone between the extremes of too much and too little.

women's hormonal profiles and overall reproductive function, promoting regular menstrual cycles, ovulation, and fertility. Even overweight women who have experienced a prior pregnancy loss and are attempting another pregnancy benefit from walking for ten or more minutes at a time—their fertility improves significantly over six months.

Get a grip on unhealthy stress. The goal is not to eliminate stress because (a) that simply isn't possible in the modern world; and (b) some stress is actually good for you. Most people rarely think of stress as a positive thing, but a form called *eustress* is just that—positive, because it motivates us, challenges us, and helps us grow psychologically, emotionally, and physically. So we want to hold on to opportunities to create that good stress at work and in our personal lives. Moderate levels of positive stress don't adversely affect reproductive function for men or women, or how long it takes a woman to get pregnant.

Instead, the goal is to minimize negative stress (aka *distress*) and/or to improve your ability to manage it. Negative stress can take a toll on reproductive health, possibly leading to hormonal abnormalities, irregular periods, and ovulation problems in women, and the reduction of sperm quality in men—especially if the stress is excessive.* You know the drill for preventing stress overload: use good time-management strategies, say no to nonessential requests, delegate responsibilities whenever possible, and develop good coping skills and a strong support network.

Social support can counteract the potentially harmful effects of stress on both your mind and body. When researchers in China examined the effects of work stress on semen quality among 384 men, they found that men with high levels of work stress had a greater chance of having swimmers classified below the WHO's threshold for "normal" sperm concentration and total sperm count than those with low work stress. No surprise there. Here's where things get interesting: the men who had

---

*   Another concern is that someone might drink too much alcohol, smoke, overeat, or engage in other unhealthy behaviors in an effort to cope with stress overload. These potentially harmful practices could adversely affect reproductive health, as well as overall health.

high work stress *and* high levels of social support had perfectly normal sperm.

In addition to seeking social support, getting a grip on stress requires finding your personal decompression valve through meditation, deep breathing, progressive muscle relaxation, yoga, or hypnosis—and using it regularly. Besides helping you get the upper hand on anxiety and worry, these practices can improve your chances of maintaining normal reproductive hormone levels. Participating in a mindfulness-based intervention or a cognitive-behavioral group program has been found to increase the chances of getting pregnant among women who are struggling with infertility. Research has found that doing diaphragmatic breathing, progressive muscle relaxation, and guided imagery, twice a day, improves sexual desire and satisfaction—factors that are often diminished by excessive stress—in healthy young adults.

Think of these steps as forms of stay-healthy insurance for your reproductive health. Combine these with measures to reduce the chemical load—and hence your exposure—in your home, as you'll see in the next chapter, and you'll enhance your health even further. It's a multilayered protection plan.

12

REDUCING THE CHEMICAL FOOTPRINT IN YOUR HOME:
Making It a Safer Haven

Knowledge can be powerful, but it can also scare the daylights out of you. If what you now know about the perilous decline in sperm counts and impaired reproductive development in men and women has you nervously wondering if you've got "enough ammunition in the arsenal" (if you're a man) or worriedly caressing your belly (if you're an expectant mother), take heart. There are several things you can do to protect your reproductive function and the reproductive health of your future child. By taking key steps to improve your lifestyle and reduce your body's burden of chemical exposures, you'll enhance your ability to preserve sperm counts, sperm motility, and your fertility whether you're a man or a woman.

In 2010, I appeared on a *60 Minutes* segment called "Phthalates: Are They Safe?" in which Lesley Stahl and I walked room to room through a suburban home, and I pointed out where phthalates were likely to be hiding. It was an illuminating experience for her and for viewers, but by focusing on phthalates we identified only a small percentage of environmental risks. Still, the room-by-room approach seemed useful, so I'm going to use it here to show you where endocrine-disrupting chemicals may be lurking in your home and how you can avoid them.

The Kitchen

It's often the hub of the home—and one of the biggest sources of exposure to phthalates, BPA, and other endocrine-disrupting chemicals. After all, these sneaky chemicals can infiltrate foods and beverages at any point in their journey from farm to fork or from manufacturing plant to cup or bottle. Want proof? When German researchers compared phthalate levels in five adults before they fasted and forty-eight hours after they fasted, a time in which they consumed only water in glass bottles, they found that levels of testosterone-lowering DEHP, and its more contemporary substitutes, in the subjects' urine dropped within twenty-four hours of the onset of the fast to just 10 to 20 percent of their initial levels. That's how quickly these sneaky chemicals can take up residence inside your body—or leave it.

To avoid numerous EDCs and other toxic chemicals in the kitchen, take the following steps:

- **Buy organic produce, whenever possible.** Sometimes it's more expensive, sometimes it's not—but if it is, it may be worth the extra investment in your health so that you can avoid ingesting trace amounts of pesticides and the inert ingredients in pesticides, which include some phthalates. If you're not inclined to buy all organic fruits and vegetables, it's smart to eliminate those that typically contain the highest pesticide residues from conventional growing methods. Every year, the Environmental Working Group (EWG, www.ewg.org), a nonprofit organization dedicated to protecting human health and the environment, releases a list of the fruits and vegetables with the highest and lowest pesticide residues, called the "Dirty Dozen" and the "Clean Fifteen," respectively. In 2019, strawberries, spinach, kale, nectarines, apples, and grapes topped the most contaminated list, while avocados, sweet corn, pineapples, sweet peas (frozen), onions, and papayas were among the least contaminated. Purchase organic fruits and vegetables whenever you can, and when you can't, rinse your produce thoroughly with tap water, then dry it with a clean towel; this will remove most of the residual chemicals. (You do not need a special produce wash.) A study by researchers at the University of California, Berkeley, found that eating organically grown food for just one week significantly reduces the levels of thirteen pesticide metabolites in the body.

• **Choose fresh, unprocessed foods.** Sticking with fresh foods— particularly fruits, vegetables, nuts and seeds, and fish—will, be- sides being more nutritious than packaged foods, help you reduce your exposure to chemicals. During processing, packaged foods come in contact with phthalates, such as DEHP and DBP—or BPA in the plastic or lining of cans—and because these chemicals aren't bound to the packaging material, they can leach into the food. Even if the label says BPA-free or phthalate-free, it may contain substitutes such as BPS and BPF for BPA or phthalate substitutes that may be as toxic as the chemicals they're replacing. It's best to try to use fewer canned and packaged foods, in general.

• **Avoid contaminants in animal products.** It's no secret that some commercially raised animals, particularly cattle and sheep, are fed hormones such as testosterone or estrogen to promote their growth or antibiotics to prevent diseases. The extent to which these hormones and drugs may affect human health when animal- based foods, including dairy products, are consumed is still hotly debated. But if you want to be on the safe side, you can look for those labeled with the USDA ORGANIC seal, which signifies that these animals have eaten only organically grown feed (without animal by-products) and weren't treated with synthetic hormones or antibiotics. Similarly, the phrases *raised without antibiotics, raised without added hormones,* or *no synthetic hormones* mean the animal received no antibiotics or hormones during its lifetime.

- **Reconsider your food-storage containers.** Phthalates and BPA are used in the manufacture of many food and beverage containers; you're exposed to these endocrine-disrupting chemicals when they seep into your foods or drinks or they're released when these containers are heated in the microwave. Plastic containers that contain phthalates have the number 3 and *V* or *PVC* in the recycling symbol. BPA is still used in many water bottles and plastic containers and in the epoxy resins that protect canned foods from contamination.* For food storage, your best bet is to use glass, metal, or ceramic containers with tops or aluminum foil. If you do opt for plastic containers, use this rhyme to help you remember which recycling codes are safer and which aren't: *4, 5, 1, and 2, all the rest are bad for you.*

- **Ban plastic from the microwave.** If you want to reheat food, don't do it in a plastic container in the microwave. Transfer it to a plate or bowl, and if you need to cover it, use parchment paper, wax paper, a white paper towel, or a domed (glass or ceramic) container that fits over the plate or bowl. Don't microwave plastic food-storage bags or plastic bags from the grocery store, even if the package is marked as safe for microwaving.

- **Prepare meals at home as often as possible.** Believe it or not, frequently dining out or getting takeout is associated with higher levels of phthalates in the body, thanks to food-packaging materials or food-handling gloves that are used. One study found that teenagers who ate out a lot had 55 percent higher levels of androgen-disrupting chemicals than their peers who only consumed food at home. Opt for home-cooked or home-assembled meals when you can.

---

* The recycling codes for plastics most likely to contain BPA are 3 (polyvinyl chloride) and 7 (polycarbonate).

- **Upgrade your cookware.** If you've been using nonstick pots and pans, it's time for a change: Nonstick cookware is made with PFOA (perfluorooctanoic acid) compounds or Teflon (a brand name for the chemical polytetrafluoroethylene). Sure, using nonstick cookware makes cleanup easier, but cooking on a heated nonstick surface gives endocrine-disrupting chemicals ample opportunity to seep into your food. If you do continue using your nonstick cookware, only use it for short periods of time at medium-low heat and discard the pot or pan if the surface becomes scratched or starts to give off flakes. In my home, we have switched to cast-iron pots and pans, which we love. Stainless steel is another good alternative.

- **Filter your drinking water.** Even if you like the taste of your tap water and trust your water supplier, it's a good idea to buy a water filter for your home (or fridge), and remember to change it regularly. As you've seen, numerous industrial and agricultural chemicals can seep into the water supply, and so can pharmaceuticals, which aren't even monitored by your water supplier. So you really don't know the full extent of what you're drinking. And drinking bottled water isn't the solution because it comes in plastic! Invest in a water treatment system for your household, whether it's an inexpensive glass (not plastic!) pitcher that you fill manually, an under-the-sink activated-charcoal or reverse-osmosis filtration system, or a whole-house carbon filter that will remove contaminants from all the water that comes into your home. (Consult NSF International, www.nsf.org, for more water-filtration-system information.) If you want a portable water bottle, get a glass or stainless steel one.

- **Clean up your cleaning products.** Carpet shampoo, all-purpose household cleaners, window- and wood-cleaning products, disinfectants, stain removers, and most other cleaning products contain potent toxins and EDCs. Go through your arsenal of household cleaning products and get rid of those that feature words such as *danger, warning, poison,* or *fatal* on the label. Replace them with products that have ingredients you can identify; here, again, the Environmental Working Group is a helpful resource (http://www.ewg.org/guides/cleaners/content/top_products). Or, you can make your own cleaning products, using water, vinegar, baking soda, or essential oils; you can find DIY cleaner recipes online.

John Darkow cartoon

Originally appeared in the *Columbia Daily Tribune,* June 17, 2008.

The Bathroom

After the kitchen, the bathroom may present the biggest opportunity in your home for exposure to EDCs and other potentially harmful chemicals. This is largely because of the cosmetics and other personal-care products we use, but other issues come into play as well. Unfortunately, the cosmetic and beauty industry is poorly regulated, and many companies have label language or brand names that suggest the products are pure, natural, fresh, or otherwise wholesome. But these terms mean literally *nothing* from a legal or regulatory stand-point.

This is especially true because the Food and Drug Administration (FDA) has far less authority over the cosmetics industry than it does the drug industry, and neither the FDA nor any other government body approves or regulates cosmetic products before they hit store shelves. Instead, cosmetic companies are responsible for substantiating the safety of their products and making sure they're labeled properly before they come on the market. All of which means, the onus is on consumers to make smart, safe (or at least less harmful) choices. To avoid numerous EDCs and other toxic chemicals in your bathrooms, take the following steps:

- **Pay attention to the labels on personal-care products.** Sometimes what you'll see is pure marketing-speak, but some phrases can be meaningful. Products that carry the USDA ORGANIC seal, for example, must contain at least 95 percent organically pro-duced ingredients—meaning, they've been grown without con-ventional pesticides, herbicides, petroleum-based fertilizers, or genetically modified organisms; the 100 PERCENT ORGANIC label indicates that a product *only* contains organically produced ingre-dients. Sometimes what a product doesn't contain is trumpeted just as loudly—and this can be worth noting. Some examples: *Fragrance-free* means no perfumes or fragrances have been added to the cosmetic or toiletry; instead, essential oils or botanical extracts that have scents may have been used to mask the smell of the basic ingredients. Similarly, *paraben-free* and *phthalate-free* indicate that these chemicals aren't in the product. Avoid cleansers and skin-care products that are labeled *antibacterial*; regular soap and water are all you need to get clean. Remember, too, a personal-care product that's supposedly free of these bad actors can lose its integrity—its phthalate-free and BPA-free status—if it's in a plastic jar or bottle, so choose products in glass whenever possible.

- **Scan product ingredients lists.** Admittedly, it may feel as though you need a chemistry degree to decipher what's in the products you're slathering on your skin, hair, or body. But you can make a modicum of sense of their ingredients lists. In particular, avoid products that contain the following EDCs or other harmful chemicals: triclosan (often in liquid soap and toothpaste), dibutyl phthalate or DBP (in hair spray and nail products), and parabens such as methyl-, ethyl-, propyl-, isopropyl-, butyl- and isobutyl-paraben (preservatives found in shampoos, conditioners, facial and skin cleansers, moisturizers, deodorants, sunscreens, toothpastes, and makeup). To closely vet the personal-care products you like, check out the Environmental Working Group's "Skin Deep" database for details. Taking these selective steps can make a difference: a study found that when teenage girls switched to using personal-care products that were labeled as free of phthalates, parabens, triclosan, and benzophenone-3 (an organic compound often found in sunscreens), their urinary concentrations of these potentially endocrine-disrupting chemicals decreased by 27 to 44 percent—in just three days!

- **Dispose of unused medications properly.** Don't flush them down the toilet; instead, mix them with coffee grounds or kitty litter,* put them in a sealed plastic bag, and place it in the trash. An even better approach is to participate in a Take Back program: check the Drug Enforcement Administration's National Prescription Drug Take Back initiative's website (https://www.dea diversion.usdoj.gov/drug_disposal/takeback/) for twice-per-year collection dates and locations near you. At other times of the year, you may be able to find an independent pharmacy in your area that

---

* The rationale is that mixing unused pills with these substances makes them less appealing to children and pets and (hopefully) unrecognizable to someone who might rummage through the trash looking for drugs.

disposes of medications through the Dispose My Meds program (https://disposemymeds.org/).

- **Ditch the vinyl shower curtain.** You know that new shower curtain smell that comes with a fresh vinyl curtain or liner? It's a result of chemical off-gassing, a release of volatile organic compounds and phthalates into the air. It's not good for you. So choose an eco-friendly option made from cotton, linen, or hemp instead.

- **Banish air fresheners.** Whether you've been using a plug-in product, a wick, or a spray air freshener, stop. All of these contain phthalates and other potentially harmful chemicals. To improve air aroma in the bathroom, use an exhaust fan, open a window, or leave an open box of baking soda in the room to absorb bad odors. Also, stick with nontoxic cleaning products for the bathroom.

Elsewhere in Your Home

A variety of chemicals may have taken up residence in other areas of your home, including your bedroom, living room, and closets. The primary offenders include phthalates, flame retardants (PBDEs), and PCBs (which still reside in many homes even though they're no longer manufactured). No one expects you to do a top-to-basement makeover of your home; that would be cost prohibitive. But you can reduce the chemical load in your home considerably. Here's how:

- **Remove wall-to-wall carpet.** Synthetic carpets, such as those made from nylon or polypropylene, can emit harmful chemicals into the air (another example of off-gassing) for many years. Natural hardwood and ceramic tile are better choices because they are the least likely to absorb dust and toxic chemicals. If you want to add an area rug, choose wool or natural plant materials such as jute or sisal. Avoid using a pad that contains PBDEs; choose a wool or felt one instead. Also, steer clear of carpets or pads that have had water- or stain-proofing treatments, which add harmful chemicals to the equation. Vacuum all carpets thoroughly, using a machine with a HEPA filter, at least once a week.

- **Prevent dust buildup.** Besides being an allergen and an unsightly nuisance, household dust can absorb and become a repository for toxic chemicals. There's no need to become an obsessive neat freak, but it's wise to elevate your dusting efforts, in particular, because household dust contains toxic chemicals from products in your home. A 2017 study found that forty-five potentially harmful chemicals—including phthalates, phenols, replacement flame retardants, and perfluoroalkyl substances (PFASs)—were found in dust in 90 percent of households sampled throughout the United States. So use a damp mop on wood or ceramic floors. Wipe furniture, windowsills, doorway moldings, and ceiling fans with a microfiber or damp cotton cloth because they hold dust particles better than others (or dry ones) do. Dust electronic equipment, including TVs, frequently because they're a common source of flame retardants. Open the windows and doors while you're cleaning, and wash your hands thoroughly after dusting and cleaning.

- **Upgrade your replacement purchases.** If you're in the market for a new stereo or media system, choose electronics without PBDEs or other brominated flame retardants. If you're ready to buy a new couch, comfy chair, or mattress, choose those that are free of flame-retardant chemicals, toxic adhesives (such as those containing formaldehyde), or plastics. (If you can't replace older foam products that have ripped covers, consider getting a cotton or linen slipcover to keep the surface intact.) Choose natural-wood

tables and cabinets that are made without synthetic wood or parti-
cleboard. And buy an organic-cotton mattress pad, not one of the
plastic barriers that will release their own chemicals into the air.

- **Leave your shoes at the door.** Besides tracking in dirt from outside, the soles of your shoes can bring in not only germs, but also
  heavy metals from soil, and pesticide residues. Research has found
  that people and pets can bring weed killers and other pesticides
  that have been applied to lawns into homes up to a week after
  treatment. Consider having a dedicated pair of "indoor shoes"
  or slippers. And wipe off your pet's paws when he or she comes
  inside.

- **Clean out your closets.** Get rid of mothballs, which contain the
  toxic chemicals naphthalene or paradichlorobenzene; to protect
  your clothes from moths, use cedar chips or blocks or lavender
  sachets in your closet. If possible, choose "green" dry-cleaning
  services or ones that use liquid carbon dioxide or wet cleaning
  methods; otherwise, air out dry-cleaned clothes by removing the
  plastic and letting them hang for a day in your garage or on a
  porch before putting them in your closet.

- **Say no to plastic bags.** Invest in reusable cloth or canvas bags of
  various sizes and carry them with you or keep them in your car
  for shopping. Wash these regularly to keep them clean.

The Playroom

If you have young kids, be aware that toxic chemicals can be present in
toys and other children's products. Even though several phthalates are
banned in concentrations above 0.1 percent in children's toys and teeth-
ing devices in the United States and the European Union, toys that are

imported from other parts of the world often contain them. Children are especially vulnerable to the effects of endocrine-disrupting chemicals since their bodies are still developing; plus, because their bodies are small, per pound of body weight they absorb more contaminants through their lungs, digestive tracts, and skin than adults do. And because young kids often put toys in their mouths, this can increase their exposure even more.

Your best bet is to scrutinize your choices when purchasing toys or kids' activities. When buying plastic toys, look for those that are labeled *phthalate-free* and *PVC-free*. Similarly, buy baby bottles and sippy cups that are labeled *BPA-free*. (Alert: this won't eliminate the BPA look-alikes such as BPS and BPF.) When furnishing the playroom, include natural materials whenever possible. Choose wooden tables and chairs, with cushions if desired, and baskets, rather than plastic bins, to hold toys and art supplies. Keep in mind: cotton fabrics and rugs are easy to clean and resist mold and mildew.

Your Yard

If you live in a house, you'll want to pay attention to the potential chemical consequences in the great outdoors—as in, your lawn or garden—even if you don't have a green thumb. That means, avoid using synthetic pesticides, herbicides, and fertilizers. They're a hazard to kids and pets and to the rest of us. If you're desperate to get rid of weeds, do it safely—by pulling them out at the roots, applying vinegar or salt to them, or by using a thick layer of mulch (such as cedar mulch or bark chips) to inhibit weed growth. Share your planet-friendly efforts and encourage others to follow your example by posting a PESTICIDE-FREE sign on your grass or in your garden. Consider, too, that your old PVC garden hose may be delivering a heavy dose of lead, BPA, and phthalates, along with the water. It may be time for a replacement. Look for a PVC-free hose that's labeled *phthalate-free* or, even better, *drinking-water safe*.

*   *   *

These are among the most common culprits that can have insidious effects on sperm counts and other aspects of reproductive health for men and women. Given the expense involved, you may not be ready to get rid of carpets, couches, cookware, and other household items that contain some of these offensive chemicals, but when you're in the market for new ones, look for items that are free of phthalates, PFOAs, flame retardants, and other potentially toxic chemicals. In the meantime, get rid of mothballs, air fresheners, scented candles, antibacterial soaps, and other items that may pose a threat to sperm and your overall health. The Silent Spring Institute offers a free smartphone app called Detox Me, which provides simple, evidence-based tips on how to reduce exposure to these chemicals in your home, and a Detox Me Action Kit, a urine test that allows you to detect the presence of common household toxins in your body. Also, make it a point to avoid handling receipts, because most of these contain bisphenol A, which can be absorbed into your body.

Think of these important steps as ways of committing to clean living, both inside and out. By improving your lifestyle habits, including your dietary choices and food-preparation techniques, and by purging your home of harmful chemicals, you'll be taking smart precautionary measures to protect your reproductive system and your overall health. As you've seen, it's possible to reduce the chemical burden that's placed on your body. But it requires diligent efforts to learn to bob and weave through the minefield of disruptive chemical influences in our midst. This is your opportunity to protect your future and your family's.

13

ENVISIONING A HEALTHIER FUTURE:
What Needs to Be Done

In 1898, UK factory inspector Lucy Deane warned of the harmful effects of exposure to asbestos dust, but her written report was largely ignored. More than a decade later, in 1911, experiments on rats raised "reasonable grounds" for suspicion that exposure to asbestos dust is harmful to the health of living creatures. Between 1935 and 1949, an alarming number of lung cancer cases were reported among asbestos-manufacturing workers, and in 1955 research established a high risk for lung cancer among asbestos workers in Rochdale, in the UK. Between 1959 and 1964, mesothelioma cancer, which affects the tissue lining the lungs, was found to be a significant problem in asbestos-manufacturing workers and people living in neighborhoods near factories that handled asbestos in South Africa, the UK, and the United States.

Nevertheless, it took until 1973 for all forms of asbestos to be recognized as carcinogenic to humans and until 1999 for many countries in Western Europe to ban the use of all types of asbestos. *That's an entire century!* But here's the real kicker: unbeknownst to many people, the United States still permits some use of asbestos, and in some developing countries (such as India), the asbestos industry continues to boom. Even after major scientific and regulatory efforts, after more than fifty years have passed, we still haven't gotten this known carcinogen out of our environment.

This story has little to do with reproductive health and plenty to do with respiratory health. But it's a powerful example of just how long it can take and how difficult it can be for important protective measures to be implemented.

Given the approximately eighty-five thousand chemicals that have been produced for use in commerce and the small number that have been tested for safety, let alone regulated, we need a better—meaning, less time-consuming and less costly—way to identify and limit exposure to risky chemicals. As an example, consider the testosterone-lowering phthalate di-2-ethylhexyl phthalate (DEHP). In 2000, John Brock, PhD, an environmental chemist, told me about a new effort at the CDC to measure phthalates in a sample of US residents for the first time. When he suggested I study them, my reaction was "What are phthalates?" He told me about some convincing studies showing that these "everywhere chemicals" were wreaking havoc on the genital tracts of male rats. Fast-forward to 2005, when my colleagues and I published our study that showed that an expectant mother with higher DEHP levels in early pregnancy was more likely to have a son with less "male-typical" genitals—for example, a shorter anogenital distance (AGD) and a smaller penis. This study and those that followed have taken twenty years and cost more than $10 million federal dollars, but they've led to important public-health action. The risk of the phthalate syndrome was believed to be so plausible that DEHP and two other testosterone-lowering phthalates were banned in toys and sippy cups in 2008.

Because of this law and the public's concerns about these health risks, levels of the "traditional" phthalates in people's bodies dropped dramatically in the United States. The pregnant women we recruited into our 2010 study had DEHP levels that were only 50 percent of those measured in pregnant moms in 2000. This was definitely a positive sign. But sadly, other phthalates have been introduced as replacements for DEHP and similarly problematic phthalates. One of these was diisononyl phthalate

(DINP). In a Swedish study, expectant mothers who had higher levels of this new substitute phthalate in their urine were more likely to have a boy with a shorter AGD than women with low levels. So swapping DINP for DEHP had not solved the problem at all, which is incredibly frustrating.

Let's pause for a moment and give manufacturers the benefit of the doubt. Let's imagine that maybe they didn't know that DINP was just as harmful as DEHP. Shouldn't they have done their due diligence and investigated the effects of this substitution before it was made? And shouldn't manufacturers have pulled this chemical from production as soon as it was discovered to be harmful? As you can probably guess, I would answer both questions with a resounding yes! But the worlds of chemistry and commerce don't always work this way. So far, this issue has been a victim of the politics of inattention, in which manufacturers have largely shirked responsibility for ensuring the safety of the chemicals in their products—and our regulatory system has allowed this to happen.

As you undoubtedly know, some people profoundly distrust the safety of vaccines or fluoride in the water supply. I keep wondering, Where are the people who are upset about the presence of harmful EDCs in everyday products? *Where is the outrage on this issue?!*

Frankly, I continue to be astonished that more public-health experts and regular citizens aren't more upset about these harmful substances. It would undoubtedly help if they were, because several things need to be done to significantly lessen the planetary burden of EDCs and make our future healthier. We need to design safer chemicals that don't interfere with the body's endocrine system, and we need to adopt testing methods—including those that will identify detrimental effects of low doses and mixtures of chemicals—that will protect us against EDCs. It's essential that regulators stop exempting ("grandfathering in") chemicals that have been used for a long time. The goal of regulatory action must shift from damage control *after* a problem has been identified to anticipating risks *before* they occur and allowing chemicals to enter the market on that basis. In other words, we need to stop using each other and our unborn children as guinea pigs for EDC exposures. And we need legislation that requires industries and manufacturers to be held accountable for the risks that come from the chemicals they produce and release into the environment.

Revamping the Regulatory Rigmarole

In the United States, changing the regulatory mechanisms, including identifying and banning chemicals that are proven to be dangerous, is an exceedingly onerous process—and considerable harm can occur while regulatory agencies are figuring this out. Still, it's certainly worth the effort to change them, because the health, vitality, and longevity of the human race and the planet depend on it. This is one of the reasons why scientists, environmental activists, health experts, and others are increasingly calling for implementation of the "precautionary principle" in public-health and environmental decision-making.

The precautionary principle shifts regulatory action from initiating damage control *after* a problem has been identified to taking anticipatory action *before* damage can occur; this is what we need to protect public and environmental health. A consensus statement from the 1998 Wingspread Conference, which included treaty negotiators, activists, scholars, and scientists from the United States, Canada, and Europe, summarizes the principle this way: "When an activity raises threats of harm to human health or the environment, precautionary measures should be taken even if some cause and effect relationships are not fully established scientifically."

A consequence of the precautionary principle is to shift the burden of proof of safety from the public to manufacturers. It also eliminates the need to wait for scientific certainty to take protective or preventive action. In some cases, strong suspicion may be enough to prevent potentially harmful chemicals from being used in everyday products. If we were to apply the precautionary principle to endocrine-disrupting chemicals and other toxic chemicals that are likely culprits in the sperm-count decline and the impairment of male and female reproductive development, human beings would face far fewer detrimental exposures on a daily basis. What we really need is for the chemical industry to adopt its own version of the Hippocratic oath—"first, do no harm."

In the European Union, a good regulatory model called REACH (short for Registration, Evaluation, Authorization, and Restriction of Chemicals) already exists. With a policy of *no data, no market,* "the REACH Regulation places responsibility on industry to manage the risks from chemicals and to provide safety information on the substances." REACH was put in place in 2007 with the goal of providing a high level of protection for human health and the environment from risks that can be posed by chemicals. It also places the burden on companies: manufacturers are held responsible for understanding and managing the risks associated with the use of their chemicals in everyday life. In my opinion, testing chemicals for hormone-disrupting potential *before* they come to market should be required throughout the world.

Under REACH, manufacturers and importers are also required to gather information on the properties of their chemical substances and to register the information in a central database that's maintained by the European Chemicals Agency. Though it's moving more slowly than health and environmental groups had hoped, REACH *is* reducing threats to human health from chemical production in the EU. For example, the REACH Dioxin Strategy, which has the goal of reducing the presence of dioxins, furans, and PCBs in the environment, has been successful: by 2014 it had achieved a reduction of industrial emissions of these pollutants by about 80 percent.

One hope is that REACH will help eliminate the unfortunate practice known as regrettable substitution: swapping in an untested compound with the function (and risk) of a known hazard. Consider the case of BPA and its replacements. As you've read, BPA is a chemical that has been used in cash-register receipts, polycarbonate water bottles, and the linings of food cans. It was first found to mimic the female hormone estrogen when it was formulated in the 1930s, and we now know that it has adverse health effects, including increasing the risk of breast cancer, recurrent miscarriage, behavioral problems in boys, and, in male BPA-exposed factory workers, impaired semen quality.

While the European Union has banned BPA in baby bottles and is phasing it out of cash-register receipts, it is still extensively used in other products, including the lining of food and beverage cans. Scrambling to find replacement chemicals, manufacturers discovered that the easiest option was to switch to another closely related bisphenol, such as bisphenol S or bisphenol F. Problem solved, right? Not exactly, because researchers are now finding that many of these BPA replacements end up in people's urine samples around the world—and that these replacement chemicals are also hormone disruptors and convey the same risk as BPA (or perhaps an even greater risk). In other words, one harmful chemical was simply used to replace another—an unacceptable practice. When I spoke to Ninja Reineke, head of science for CHEM Trust, a leading nonprofit organization working on the EU's REACH chemical regulation, in the fall of 2019, she confirmed that despite the passage of REACH legislation, regulators are not yet controlling the use of regrettable substitutes, even in the EU.

It shouldn't be this way. As Joseph Allen, an assistant professor of exposure assessment science at Harvard University's T. H. Chan School of Public Health, wrote in a 2016 opinion piece for the *Washington Post*, "Innocent until proven guilty may be the right starting point for criminal justice, but it is disastrous chemical policy. We need to recognize regrettable substitution for what it is: repeated substitution of toxic chemicals with equally toxic chemicals in a dangerous experiment to which none of us knowingly signed on."

I agree with him a thousand percent. We are all essentially unwitting participants in a chemical game of reproductive Russian roulette because regulation of the chemical and manufacturing industries continues to operate on a "business as usual" basis, with chemicals considered safe until they're proven guilty. The chemicals I'm most concerned about are "stealth chemicals"—phthalates, BPA, fluorinated compounds, and

PBDEs—because they enter our bodies silently, secretly, and without our knowledge. Unlike drugs, which are monitored for safety by the FDA and sold with detailed warning labels, environmental chemicals are largely unregulated, and few are identified on labels.

Back to the Drawing Board

A critical step toward eliminating EDCs from our daily lives is to create newer, smarter chemicals, such as those promised by "green chemistry." This field embraces the overarching goals of developing more resource-efficient and inherently safer molecules, materials, and products. To achieve these goals, chemists must be able to assess potential hazards of the chemicals they develop. First and foremost of these goals should be avoidance of endocrine disruption.

One novel approach that looks particularly promising is known as the Tiered Protocol for Endocrine Disruption (TiPED), which applies principles and tests from the environmental health sciences to identify potential endocrine disruptors. Formulated by a team of renowned scientists from multiple disciplines, this protocol is designed to help chemists identify and avoid chemicals that are likely to disrupt the human endocrine system. Through this system, chemicals that are identified as potential EDCs can then be removed from product development or redesigned to avoid the identified mechanisms of EDC action—before these products enter the marketplace. By facilitating early identification of EDCs, TiPED's ultimate goal is to reduce the environmental and public-health risks from these chemicals.

This is certainly a step in the right direction, especially for detecting adverse effects from exposure to low doses or concentrations of these chemicals. The notion that "the dose makes the poison" is a core but outdated assumption that underlies traditional toxicology. This assumption is credited to Paracelsus, a Swiss physician, alchemist, and astrologer, who, nearly five hundred years ago, expressed this basic principle: "All things are poison, and nothing is without poison; only the dose makes a thing not a poison." His idea was that the higher the dose, the greater the (adverse) effect is, to humans and perhaps other creatures. But that isn't always true, and we need better testing protocols to tease out risks from both high *and* low doses.

As Terry Collins, PhD, a professor of green chemistry at Carnegie Mellon University, and a user and proponent of TiPED, notes, "Low-dose toxicity is much more insidious than high-dose toxicity and is the likely cause of much, if not most, of the reproductive damage we are seeing in multiple species." If we can develop more effective testing protocols and better ways to screen chemicals to protect public health, we will have a much better chance of stemming the steady decline in male and female reproductive function that's currently underway.

One of the first steps in chemical regulation is to identify the harmful effects of the chemicals in question. Much of the research showing adverse effects of EDCs and risky exposures to lead and radiation comes from animal studies, as you've seen. These early results are then typically followed by studies in humans, at a cost of millions of dollars and five to ten years of research for a single study. In the future, these studies, with both humans and animals, need to be set up in a fashion that reflects how people are actually exposed to these chemicals in real life, because the harm that's detected varies with the dose or level of a particular chemical, as well as the timing of the exposure and the combinations of exposures. We need to keep it real.

Troublesome Suppositions and Assumptions

The truth is, current testing protocols are not adequate to protect public health because they make unfounded assumptions about the nature of the risk that EDCs, in particular, pose to human health. Following the "dose makes the poison" principle, current testing begins at a high (toxic) dose and continues at lower doses until a dose is identified at which little or no risk is seen. Then, based on Paracelsus's law, it is assumed that lower exposures are safe and therefore not tested or regulated. This principle underlies most regulations in Europe and the United States, and it's intended to protect people from risks from toxic exposures. Everyone assumes that this assumption is correct—but it misses a crucial part of the picture: in some instances, exposure to low doses of certain chemicals could be just as risky or perhaps even riskier than exposure to high doses.

This can happen when a particular chemical causes different adverse effects at lower and higher doses. For example, thalidomide is a sedative and hypnotic that was used in Europe in the late 1950s and the 1960s until it was discovered that it caused limb malformations, particularly absent or shortened limbs, and that high doses could cause fetal death. If you were doing a study of limb defects in live-born babies following prenatal thalidomide exposure, and you were to plot a graph showing the risk of limb defects as a function of dose, at a high dose the risk would appear to drop. Why? Because at high doses many of the most affected fetuses will die, and those that survive will have relatively few limb defects. Obviously, this does *not* mean the drug isn't harmful to human development. Needless to say, death is a sure sign of toxicity.

In fact, evidence from decades of research that combines toxicology, developmental biology, endocrinology, and biochemistry has demonstrated that this "law" of Paracelsus's cannot be assumed for EDCs. On the contrary, some chemicals, particularly those that behave like hormones—such as the estrogenic compound BPA—may have even more harmful effects at lower doses than at higher ones.

If you make a graph, charting risk against dose, the graph according to Paracelsus would continue to climb as the dose increases; it's an example of a *monotonic curve*, meaning one that doesn't change direction. But when lower doses are riskier than higher doses, *that* line climbs with increasing dose up to a certain point and then decreases (picture an inverted *U*). These dose-response curves are examples of *non-monotonic dose-response (NMDR) curves*—a mouthful of a term but one that's good to know.

Think back to the "sweet spot" for exercise and fertility that you read about in chapter 6. As you saw, reproductive fitness increased with an increasing amount and intensity of physical activity, but after a certain point, it started to be a risk for *infertility*. Not only was there a point of diminishing returns, but also at some point reproductive harm could occur. So if we draw a curve representing how long it took someone to become pregnant, based on how much she exercises, it might look like the letter *U*—that's another NMDR curve.

In a review of 109 studies on the effects of BPA, published between 2007 and 2013, researchers found NMDR curves in more than 30 percent of the studies. This suggests that current risk-assessment methods, in which supposedly safe exposures to low doses are predicted from high-dose exposures, do *not* protect the public from potentially risky doses of BPA. In such cases it is *incorrect* to assume that lower doses are safer than higher doses—yet this assumption continues to underlie regulatory testing for environmental chemicals.

Low-dose safety cannot be deduced from high-dose testing of a particular chemical. Treatment of estrogen-dependent breast cancer by the drug tamoxifen is a good example. In studies of breast tumor cells, it was observed that while high, therapeutic concentrations of tamoxifen inhibited estrogen-stimulated proliferation of breast cancer cells, lower concentrations of the same drug actually stimulated breast cancer cell growth in cancers that were estrogen-dependent. This is a known phenomenon in cancer therapy, and it's referred to as the "tamoxifen flare."

In other words, a chemical can cause effects at low doses that don't happen at higher doses of the same chemical, or vice versa. That's why the entire approach to regulatory testing needs to be revamped to protect human health.

Beacons of Hope

Given the daunting challenges faced by regulators struggling to get testing protocols right and chemists designing chemicals that are "endocrine-disruptor-free" and "fossil-fuel-independent," it's a wonder that there has been any progress at all. But significant steps *have* been made toward more effective regulations, cleaning up our air and water, and saving many endangered species, in the process. For example, as you saw in chapter 9, the 2018 study of bird decline on farmland also included some bright spots. Due to conservation efforts, populations of wetland birds such as ducks and geese are on the rise. Encouragingly, the population of raptors such as bald eagles, which were close to extinction before the prohibition of the insecticide DDT, is increasing, too, thanks to endangered-species protections and other federal laws. Previously, DDT made the shells of their eggs so weak that, in trying to incubate them, the bald eagles would instead crush them.

By 1963 only 417 bald eagle breeding pairs remained. Following the ban on the use of DDT in 1972, the bald eagle's comeback was spectacular, with ten thousand breeding pairs currently in the lower forty-eight states. That's a reproductive victory if ever there was one. Other species can be helped as well by the adoption of sustainable agricultural practices that minimize the use of pesticides and offer farmers incentives to set aside land for wildlife.

Other species were preserved by the 1972 ban of DDT, the Endangered Species Act (ESA) of 1973, or its predecessor, the Endangered Species Preservation Act (1966). The whooping crane was another, at least partial, success that came from the ESA. Because the millinery industry prized crane feathers as decorations on ladies' hats, these birds were hunted to near extinction, and by 1941, only sixteen of these birds remained in the United States. After the ESA became law, the surviving whooping cranes were rounded up for captive breeding. Currently, a few hundred whooping cranes are back in the wild, living in several distinct breeding and migrating populations.

Despite these significant advances, we still have a long way to go, and it's critical that these species-protection efforts continue and that new ones be introduced. In 2019 the World Wildlife Fund listed forty-one endangered species (eighteen of which are critically endangered), nineteen vulnerable species, and nine near-threatened species. There is more work to be done.

Better Chemical Regulation

Almost daily, we hear encouraging news about initiatives that effectively decrease environmental pollutants in the United States and abroad. On July 1, 2020, Denmark became the first country to ban PFAS chemicals from food packaging. PFASs are used to repel grease and water in packaging for fatty and moist foods such as burgers and cakes. This is excellent news because PFASs are among the "forever chemicals" (so called because they don't break down in the environment). Another example of protective legislation: Hawaii recently passed a law that bans the chemicals oxybenzone and octinoxate in skin-care products, starting in 2021, because they are damaging to coral reefs, which are crucial to marine and human life. Thanks to legislation such as this, progress is afoot. But, again, there's more work to be done.

Not many people know of the US Consumer Product Safety Commission (CPSC), a federal agency that was created by Congress in 1972 to "protect the public against unreasonable risks of injuries and deaths associated with consumer products." The CPSC has jurisdiction over thousands of types of consumer products, and the commission has been investigating various risks from phthalates in these products. As part of that investigation the commission formed a Chronic Hazard Advisory Panel (CHAP), which examined the health effects of phthalates in children's toys and child-care items and brought in researchers whose studies looked at health risks from phthalates. In 2015, I presented results of our studies on phthalates to the commission; two years later, the CPSC determined that eight phthalates cause harmful effects to male reproductive development and banned children's toys and child-care items that contained more than a minimal amount of these phthalates (0.1

percent). A short-term ban was already in place on three phthalates in children's toys and child-care products, thanks to the Consumer Product Safety Improvement Act of 2008; the 2017 ruling made that ban permanent and expanded it. And yet . . . other, newer hormone-altering phthalates remain on the market.

Around the world, different countries are stepping up their efforts to limit environmental damage and reduce humans' exposure to EDCs. Costa Rica, which is among the top five countries leading the way with renewable resources, has made a commitment to become single-use plastic-free and to derive all its energy from non-fossil-fuel sources by 2021. Pakistan has moved to ban single-use plastic bags. And Australia has come up with a way to decrease the entrance of plastic and other garbage into the oceans: the city of Kwinana recently installed a filtration system on the outlet of drainage pipes that catches all the large debris, thus protecting the environment from garbage and plastic contamination; when the net is full, it is picked up and emptied into special trucks. Having less plastic in the environment will automatically reduce the presence of some EDCs that could imperil reproductive health for all living creatures.

Some businesses and retailers are also helping to reduce consumers' exposure to harmful chemicals. For example, Wegmans is a grocery store chain that I loved shopping at when I lived in Rochester, New York, from 2005 to 2010. When the store's management read in the local paper about the work I did on phthalates while I was at the University of Rochester, they asked me to talk to their buyers about phthalate-containing products; after I met with them, the store flagged phthalate-free products on their shelves so consumers could find them easily.

Interestingly, Walmart, the largest discount retailer in the world, has developed lists of chemicals it wants phased out of products they carry, lists they share with suppliers. Unbeknownst to many, Walmart supports a large sustainability program focused on three areas: food waste, deforestation, and reducing plastic waste. Recently, the Home

Depot, the largest home-improvement retailer in the world, announced that it will no longer sell carpets and rugs that are treated with perfluoroalkyl and polyfluoroalkyl substances in Canada, the United States, and online.

An increasing number of eco-friendly manufacturers have contributed to the worldwide effort to reduce EDCs and other toxins in our daily lives, sometimes under the umbrella of "corporate social responsibility" (CSR). One of the earliest and most effective promoters of CSR is Patagonia, a company that since 1973 has specialized in outdoor apparel and is still owned by its founders, the noted mountain climber Yvon Chouinard and his wife, Malinda. For most of the company's existence, it has been pioneering efforts to steer the clothing industry in a more sustainable direction. In 2010, Patagonia helped found the Sustainable Apparel Coalition, an alliance of companies from the clothing and footwear industries whose members are working to make more sustainable decisions when sourcing materials and developing products.

The point: It has become increasingly apparent that there is social, economic, and environmental value in investing in sustainability. Each year at the World Economic Forum in Davos, Switzerland, the world's most sustainable companies (the Global 100) are chosen from a list of about seventy-five hundred companies, all of which generate more than $1 billion in annual revenue. This list ranks corporations on their performance in reducing carbon and waste production, their gender diversity among leadership, revenues derived from clean products, and overall sustainability. An increasing number of global companies are recognizing that incorporating sustainability into corporate values is good business. What's needed now is a broader recognition that "sustainability" must include the development of products that are *not toxic, not hormonally active, and not bioaccumulative* (meaning, they don't accumulate in the tissues of a living organism). We, as consumers, should support sustainable-product development and sustainable investing by companies with our spending habits.

It's true that human beings created these toxic chemicals and unleashed them into the world. We also have the power to mitigate or reverse them. While we've started to make progress on this front, we need more initiatives like the ones you just read about, and we need them to be implemented faster. It should be the government's responsibility to require premarket testing of these chemicals and to monitor companies' compliance. (Right now, the onus is on us as consumers to take the right steps to protect ourselves, but it shouldn't be.) We need people around the world to cast their votes for leaders who will make it a priority to ban harmful chemicals and industrial practices that poison the planet.

The status quo has persisted for too long—and it's endangering the reproductive health and survival of human beings and other species. The time to correct course is overdue and more important now than ever. I see this as both a scientific and a moral imperative, because otherwise we and other species could end up marching toward the brink of extinction or obsolescence.

CONCLUSION

As science fiction writer Isaac Asimov noted, "The saddest aspect of life right now is that science gathers knowledge faster than society gathers wisdom." Admittedly, he wasn't talking about EDCs, lifestyle habits, or reproductive health—but the quote is certainly relevant to these issues. As you've seen, numerous damaging forces are contributing to the dramatic decline in sperm counts in Western countries and the alarming rise in reproductive health problems in men and women. Many of these trends have occurred at approximately the same rate—1 percent per year—which can hardly be a coincidence.

We're not alone; these influences have also been poisoning other species and the ecosystems that we share. As a species, we're failing to propagate and repopulate ourselves, and we're hindering the ability of other species to do so. We're increasingly recognizing these realities (the "knowledge" Asimov alluded to), but we haven't gathered the wisdom necessary to make changes that would put our futures back on a healthier course.

Consider this book a rallying cry for raising awareness about these issues. My hope is that you now feel inspired to take notice of the potentially harmful lifestyle and environmental influences in our modern world and to take action in whatever ways you can to reverse, reduce, or counteract these damaging effects. We can no longer afford to behave as though it's business as usual. The canary has sung, loudly, clearly, and shrilly; now, it's up to us to heed the message and take steps to protect our legacies.

We need to upgrade our health habits and become more mindful about the items we choose to use or bring into our homes or workplaces. Problems with sperm count and quality can sometimes be turned around when men improve their lifestyle habits or reduce their exposures to toxic environmental influences, as you've seen. While women don't have as much of an opportunity to hit the reset button on their reproductive health, they can sometimes improve the regularity of their menstrual cycles and ovulatory patterns and enhance their fertility with their eating and exercise habits, in particular. And, of course, women can play a tremendous role in safeguarding their babies in utero, which can have positive effects for subsequent generations.

We should also concentrate on cleaning up the messes we've made in various ecosystems. Species are interdependent, so reversing damage to one habitat can have a positive trickle-up effect from one species to another. A case in point: In the fall of 2019, a report emerged about coral gardeners slowly restoring Jamaica's "undersea rain forests" and the dazzling diversity of life they shelter. As the *Washington Post* reported, "After several natural and man-made disasters in the 1980s and 1990s, Jamaica lost 85 percent of its once-bountiful reefs. Meanwhile, fish catches declined to a sixth of what they'd been in the 1950s, pushing families depending on seafood closer to poverty." Now the coral and various species of tropical fish are gradually reappearing, thanks to the conscientious efforts of humans. As Stuart Sandin, a marine biologist at the Scripps Institution of Oceanography in La Jolla, California, said, "When you give nature a chance, she can repair herself."

Personally, I believe the same is true of human beings.

It's a mistake to underestimate the power of human ingenuity. Humans are remarkably resilient and resourceful creatures when we set our minds on the right goals. We've accomplished amazing turnarounds in the past—eradication of smallpox and polio in the United States, improvement in air quality throughout the country since passage of the Clean Air Act in 1970, and the successful cleanup and environmental restoration of the Great Lakes region's most heavily polluted areas since the 1980s. Between 1976 and 1991, lead levels in human blood dropped by 78 percent, mostly because 99.8 percent of lead was removed from gasoline and lead was eliminated from soldered cans. I believe we can achieve similarly remarkable reversals when it comes to the effects of EDCs on reproductive health.

To take the next steps that are needed throughout the United States and the world, we need to share information about the dangers of endocrine-disrupting chemicals and why it's important to get them out of our environment. Surprisingly, when I ask, even at scientific meetings, how many people know about endocrine disruption, the number of hands that are raised is still discouragingly small. This information can and should be made part of middle and high school science programs, as well as medical school curricula. That type of knowledge dissemination will make it more likely that at some point, in the future, your physician will routinely provide you with up-to-date recommendations on products and practices that are found to be risky and ways to assess the safety of your environment.

We need to increase awareness of the importance of reproductive health—for our own sake, our offspring's, and the health of the planet. Sadly, reproductive health is the poor stepchild of medical research. The National Institutes of Health has twenty-seven institutes that fund studies on a wide range of diseases—cancer, diabetes, allergy and infectious disease, dental and craniofacial diseases, and even aging—but not reproduction. The closest the NIH comes is with the National Institute of Child Health and Human Development, which supports research on birth defects and maternal mortality but not sperm decline.

Despite these gaps in research, knowledge, and action, I do think it's possible for us to fix what threatens and endangers human life, and here's why: We've made huge strides in understanding how exposure to everyday products can damage our hormonal systems. We now understand the exquisite sensitivity of the fetus, something that was undreamed of when the fetus was believed to be protected by the placenta and the womb. We know that all of us, including newborn infants, are continuously exposed to more than one hundred chemicals, which have the ability to profoundly alter basic biology. And we know that the archaic beliefs underlying much of chemical regulation don't protect us. Scientific skepticism aside, I remain cautiously optimistic about our collective future. I have to.

I've spent most of my professional life trying to figure out how our environment can interfere with such basic functions as conceiving and delivering a healthy child—and how we can protect ourselves. Unfortunately, in the past, after writing and speaking to other scientists about my research results, I've felt that the people who could make a difference still weren't hearing my message. True, the (unexpected) tsunami of interest in my colleagues' and my 2017 sperm-decline meta-analysis was encouraging. It felt to me that finally scientists, journalists, and the public were taking this threat seriously. But even a huge number of hits and citations can quickly be forgotten, as attention moves on to the next exciting scientific finding.

The good news is, we're *finally* getting some of the answers we need to protect human reproductive health, as well as that of other species. Which is why I've written this book. It's clear that "first-generation" chemicals that were produced after World War II haven't been good for our species or the health of the planet. What the world urgently needs is a new generation of chemicals that can be used in everyday products without threatening *our* health or that of future generations, other species, and the environment at large. This is a watershed moment, a point at which we have enough data and sufficient motivation to make at least some changes that are necessary to stop that "1 percent effect" from continuing—at least at the same rapid rate.

But there are still plenty of unanswered questions. When I present the sperm-decline data, I'm often asked, How long can this go on? Is it getting better or worse? Can sperm count recover?

As a scientist and a statistician, I can't speculate, but I can look to the past for patterns. I'll be honest: right now I don't see signs of the decline leveling off. But I do think that a diminished sperm count can be restored. After all, men whose sperm were totally wiped out by DBCP went on to father children when they stopped working with the pesticide. That's encouraging evidence right there. By eliminating our exposure to other chemicals, I suspect similar reproductive recoveries can be made.

Still, the ultimate question to me is, How can we limit or prevent risky exposures from previous generations from being passed on to the developing fetuses in future generations? What people can do about their own exposures is the relatively easy part. But how we could potentially limit the intergenerational effects is the stuff of future science. My hope is that we'll eventually figure that out, too, so that we can protect the future of the human race, the planet, and our legacy, for generations to come.

ACKNOWLEDGMENTS

Sometimes it's impossible to predict what life will send in your direction— *Count Down* is an example of this, in a wonderfully surprising way. Shortly after the 2017 publication of a meta-analysis study I coauthored that showed a precipitous decline in sperm counts and sent shock waves throughout the world, I received a call from literary agent Jane von Mehren, asking if I'd consider writing a book about the dramatic decline in sperm count and the implications for human beings and the world at large. Dozens of journalists from around the world had interviewed me about the study; one was Stacey Colino, who'd interviewed me for Vice .com and was close to and highly respected by others in my professional network. When Jane and I approached Stacey about collaborating on a book, I got lucky because she said yes! We made a terrific team; Stacey kept my vision clear and focused, always gently and professionally, and helped me expand the scope of the book's content. I can't imagine a better collaborator.

When Scribner acquired *Count Down*, I got lucky again! Thank you to Nan Graham, who saw the potential in *Count Down* and enthusiastically supported it, to Daniel Loedel, who provided invaluable editorial insights during the first half of the writing process, and to Rick Horgan, who skillfully and gently guided the project to its successful completion, with the assistance of Beckett Rueda, who was incredibly helpful. Their questions helped me push past what I would have said to my usual academic audiences; I became emboldened by their interest and support. I'm also grateful to Jane von Mehren, who held my hand and provided sage advice throughout the entire process. And big thanks to Grace Martinez, the wonderful graphic designer who contributed so much to *Count Down* visually. In many ways, the book, like the science behind it, has been a collaborative effort, and it has grown richer and become more interesting as it has been influenced by each of the players.

For providing key support and advice during the book's development, I thank the Science Communication Network, especially Pete Myers and Amy Kostant and Terry Collins. You've always had my back, for which I can't thank you enough. Huge thanks to my 2017 study collaborators Hagai Levine, Anderson Martino-Andrade, Rachel Pinotti, Niels Jorgensen, Jaime Mendiola, Dan Weksler-Derri, and Irina Mindlis. I'm extremely grateful to Jeremy Grantham, Jamie Lee, and Ramsay Ravenel at the Grantham Foundation for their interest in sperm decline and support for my work and *Count Down*. Thank you to Rimjhim Dey, Andrew DeSio, and others at Dey whose work was critical to increasing the impact of *Count Down* and its message. I'm also immensely grateful to the scientists and experts whose research we relied on or who provided important insights that we incorporated in the book: Jane Muncke, Elena Rahona, Michael Eisenberg, Darrell Bricker, Ritch Savin-Williams, Michelle Ottey, Jack Drescher, Alice Domar, Marcia C. Inhorn, Sharon Covington, David Møbjerg Kristensen, Pat Hunt, Thea Edwards, Brandon Moore, Alison Carlson, Cynthia Daniels, Sheri Berenbaum, Dan Perrin, and Rick Smith. Thank you to the many people who generously shared their stories and concerns about issues of infertility, miscarriage, gender identity, and genital anomalies.

To the many researchers whose years of dedication and commitment produced much of the science in *Count Down*, I extend my thanks and gratitude—without you, this book wouldn't have happened. To Lou Guillette, Fred vom Saal, and Ana Soto, my fellow "Endocrine Disruptor

Cry Babies," as the Junk Science website dubbed us in 1995, being on the front lines with you and other "early EDC fighters" John MacLachlan, Howard Bern, and Theo Colborn was inspirational and life changing; it was what brought me into this tough, exciting, and terribly important area of research. Working with Niels Erik Skakkebaek and members of his department at the University of Copenhagen was eye-opening. Niels Erik's insights into the decline in male reproductive function and its fetal origins changed the field and my personal research path.

To John Brock, who first suggested I study phthalates, and Antonia Calafat at the CDC—you and other brilliant environmental chemists generated the data that have made it possible to understand the extent to which the fetus is exposed to environmental chemicals. Without you, we'd still be looking at the fetal environment as a "black box." My thanks also extend to environmental toxicologists including Jerry Heindel, Earl Gray, Paul Foster, and David Kristensen, who showed us what havoc these environmental chemicals can cause in mammalian hormonal systems and how they do that. Without you, we epidemiologists would be limited to reporting observations, without any insights into mechanisms or biological plausibility.

To the researchers who served as center directors of my cohort studies (TIDES and SFF)—Bruce Redmon, Amy Sparks, Christina Wang, Erma Drobnis, Sheela Sathyanarayana, Emily Barrett, Nicole Bush, and Ruby Nguyen—and the thousands of families who participated in these studies, receiving little compensation, thank you for hanging in there with us and making these studies possible. To Ruthann Rudel (Silent Spring Institute), Ninja Reineke (CHEM Trust), Ken Cook (EWG), and others, whose work provides an invaluable link between the science and action, thank you for all you do.

And finally, thank you to my patient and long-suffering husband, Steven, who was always available to listen and help, while tolerating my high levels of anxiety and even at times panic while working on this book. This has been such an exciting journey for me, and I've learned so much from it! Thank you to everyone who helped make *Count Down* happen.

RESOURCES

Because Health: A nonprofit environmental health site that offers science-based tips and guides for buying safer cookware and dishes, uncontaminated foods, and nontoxic personal-care products—to help people live more healthfully. www.becausehealth.org/

Breast Cancer Prevention Partners: An organization that provides information for reducing toxic exposures—in food packaging, cosmetics, and other everyday products—to protect people's breast health and reproductive health. www.bcpp.org/

CHEM Trust: A website that offers terrific fact sheets about hazardous chemicals and their impacts on health, as well as news about chemical legislation in Europe (such as REACH). chemtrust.org/

Environmental Defense Fund: A leading global nonprofit organization that promotes research related to preserving the health of both the environment and its populations, including humans. www.edf.org/

Environmental Health News: A publication of Environmental Health Sciences, a nonprofit organization dedicated to environmental health issues, including climate change, the plastic pollution crisis, and harmful chemicals such as BPA. www.ehn.org/

Environmental Working Group: A nonprofit dedicated to protecting human health and the environment. The group offers excellent Shopper's Guides with scientifically based recommendations on choosing healthy consumer products (from cosmetics to cleaning products) and uncontaminated foods (including produce). https://www.ewg.org/

The organization also offers a Healthy Living App (www.ewg.org/apps/) with ratings of more than 120,000 food and personal-care products, to help consumers make the healthiest choices.

Made Safe: A program that certifies safe brands of cosmetics, household products, apparel, bedding, and other products, after a rigorous screening and evaluation of ingredients and materials. Check out their new Healthy Pregnancy Guide. www.madesafe.org/

National Resources Defense Council: An organization that works to safeguard the earth, including its air, water, people, plants, and animals, from pollution, chemicals, and other toxic effects. www.nrdc.org/

Program on Reproductive Health and the Environment: Under the auspices of the University of California, San Francisco, this program offers valuable resources that can help minimize people's exposure to reproductive toxins in everyday life. prhe.ucsf.edu/

Safer Chemicals, Healthy Families: A coalition of organizations and businesses working to safeguard families from toxic chemicals in our homes, workplaces, schools, and in the products we use. saferchemicals.org/

Safer Made: An organization that invests in companies and technologies that eliminate the use of hazardous chemicals in consumer products and supply chains. Its newsletters highlight developments in phasing out certain chemicals and progress on other environmental issues. www.safermade.net/

Silent Spring Institute: A scientific research organization dedicated to uncovering the links between environmental chemicals and human health. silentspring.org/
On the preventive front, the organization developed Detox Me (silentspring .org/project/detox-me-mobile-app), a free mobile app to help consumers reduce their exposure to toxic chemicals in their everyday surroundings.

Toxic Free Future: An organization that conducts original research on the complex science underlying different aspects of environmental health and advocates for the use of safer products, chemicals, and practices to ensure a healthier future. toxicfreefuture.org/

If you'd like to read more about the harmful environmental exposures discussed in *Count Down,* **I recommend:**

Silent Spring (1962) by Rachel Carson explores the damage inflicted by synthetic pesticides, especially DDT, not only on insects but also on bird and fish populations and even children. This revolutionary book kick-started the environmental movement and led to a ban on DDT.

Our Stolen Future: Are We Threatening Our Fertility, Intelligence, and Survival? A Scientific Detective Story (1996) by Theo Colborn, Dianne Dumanoski, and John Peterson Myers is also a classic in the field. It ultimately influenced government policy and helped foster the development of a research and regulatory agenda within the US Environmental Protection Agency.

Slow Death by Rubber Duck: How the Toxic Chemistry of Everyday Life Affects Our Health (2009) by Rick Smith and Bruce Lourie offers a down-to-earth and often amusing look at how everyday living creates a chemical soup inside each of us—and what we can do to minimize our exposures.

Better Safe Than Sorry: How Consumers Navigate Exposure to Everyday Toxics (2018) by Norah MacKendrick provides insights into the chemical exposures we face daily, the policies and regulations that surround them, and how consumers can try to avoid them.

The Obesogen Effect: Why We Eat Less and Exercise More but Still Struggle to Lose Weight (2018) by Bruce Blumberg is about obesogens, chemicals that disrupt our hormonal systems and alter how we create and store body fat. The book explores how these chemicals work, where they are found, and practical steps we can take to reduce our exposure.

GLOSSARY

androgens: Hormones that are essential for growth and reproduction in both men and women. Testosterone (the primary androgen) is produced in the testes in men and (at much lower levels) by the ovaries in women. (A chemical that is antiandrogenic lowers androgen, usually testosterone.)

anogenital distance (AGD): The distance from the anus to the genitals; it's a marker of how much androgen an infant was exposed to during early pregnancy. It is usually 50 to 100 percent longer in males than females. Teens and young men often refer to it with slang terms, such as *gooch* or *taint.*

anti-Mullerian hormone (AMH): A hormone produced in the female by ovarian follicles. In a mature woman AMH reflects ovarian reserve and can be used as a marker of PCOS. Early in a pregnancy, a male fetus's testicles produce it, and AMH stops the development of structures that would otherwise become the ovaries, uterus, and upper vagina.

assisted reproductive technology (ART): A term that refers broadly to all medical technology used to achieve pregnancy, including fertility medications, in vitro fertilization, and surrogacy.

autism spectrum disorders (ASD): A group of developmental disorders, including autism and Asperger's, that can cause significant social, communication, and behavioral challenges.

azoospermia: A condition where a man's ejaculate is completely devoid of sperm—as in, none, zilch, nada, no sperm at all.

bisphenol A (BPA): A chemical that's added to polycarbonate plastics to make them lightweight, clear, and hard (think water bottles). It's also found in the lining of canned-food containers, cash-register receipts, and pizza boxes. Most important, BPA is an endocrine disruptor that mimics the hormone estrogen.

cannabidiol (CBD): One of more than a hundred cannabinoids (compounds) in cannabis (marijuana). CBD is mildly psychoactive, but CBD alone (without THC) is nonintoxicating and won't cause a high.

cisgender: A term for people whose gender identity matches the sex they were assigned at birth.

congenital adrenal hyperplasia (CAH): A group of genetic disorders that lead to a decrease in the hormone cortisol and an increase in male sex hormones (androgens) in both sexes. In girls this can lead to masculinization of the genitals and more male-typical play.

cortisol: A steroid hormone that helps the body respond to stress—it's one of the "stress hormones." Cortisol is released during stressful times as part of the body's fight-or-flight response, to give the body an energy boost.

cryobanking: Storing cells (such as egg and sperm), tissues, or organs at low or freezing temperatures to save them for future use. Also called cryopreservation.

cryptorchidism: Undescended testicles, a male birth defect that's usually minor. (The testicles can move up and down in the scrotum, and the location often changes during the first year of life.)

deoxyribonucleic acid (DNA): A large molecule (aka macromolecule) found in the chromosomes of almost all organisms. DNA contains the instructions an organism needs to develop, live, and reproduce.

desistance (or desistence): A term referring to the phenomenon whereby people with gender dysphoria ultimately decide not to transition their gender identity. In the field of criminology, the term describes the cessation of offensive or other antisocial behavior.

dibromochloropropane (DBCP): Used in the past as a soil fumigant and pesticide, this chemical was banned in the United States in the 1970s when it was discovered that it caused azoospermia (absence of sperm) in exposed workers.

dibutyl phthalate (DBP): A chemical commonly used in polyvinyl chloride (PVC) and found in many home and personal-care products. It is an endocrine disruptor and one of the more potent antiandrogenic (testosterone-lowering) phthalates.

dichloro-diphenyl-trichloroethane (DDT): Developed in the 1940s, DDT was the first modern insecticide to control insect-borne human diseases (such as malaria). Its widespread use led to DDT resistance and adverse environmental and human health effects. Rachel Carson's exposé of these risks in *Silent Spring* led to severe restrictions on its use.

diminished ovarian reserve (DOR): A condition in which the number and quality of a woman's eggs is lower than expected for her biological age. Also called premature ovarian aging (POA) and premature ovarian failure (POF).

disorders of sex development (DSD): Previously called **intersex**, DSD includes a range of conditions that lead to abnormal development of the sex organs and ambiguous genitalia—meaning, genitals that are not clearly male or female.

di-2-ethylhexyl phthalate (DEHP): Like DBP, DEHP is an endocrine disruptor and one of the more potent antiandrogenic phthalates. It, too, makes plastic soft and flexible and is found in foods, food containers, and a wide range of household products.

endocrine-disrupting chemicals (EDCs): Chemicals, usually man-made, that mimic, block, or interfere with hormones in the body's endocrine system.

endometriosis: A disorder in which the tissue that makes up the lining of the womb (uterus) grows outside the uterus. This can lead to subfertility as well as painful periods and sexual intercourse.

epigenetic changes: *Epigenetics* literally means "above" or "on top of" genetics. Epigenetic changes refer to external changes to DNA that turn genes "on" or "off." These changes don't alter the DNA sequence itself, but instead, they affect how cells "read" genes.

erectile dysfunction (ED): Often called impotence, ED is the inability to get and keep an erection that's firm enough for intercourse.

estrogens: Estrogens (estrone, estradiol, and estriol) are hormones produced primarily by the ovaries in women. While estrogen is thought of as the "female hormone," it's made at much lower levels by the adrenal glands and testes in men.

follicle-stimulating hormone (FSH): The hormone responsible for the growth of ovarian follicles in women. In men, FSH plays a role in sperm production.

gender dysphoria: The feeling that one's emotional and psychological identity as male or female is out of sync with one's biological sex.

gender nonconforming: A term that means the person's gender expression does not correspond to traditional notions of masculinity or femininity.

human chorionic gonadotropin (hCG): A hormone that's important in the early stages of pregnancy, when it is produced by the cells that surround the growing embryo. It can be detected as early as one week after fertilization. Low levels of hCG are also produced by the pituitary gland in men and nonpregnant women.

hypospadias: A (rare) male genital birth defect in which the urine tube opens on the underside (instead of at the end) of the penis. (It's part of the testicular dysgenesis syndrome.)

infertility: Not being able to get pregnant after one year of unprotected sex. (Confusingly, it's not simply the opposite of fertility, which is the capacity to conceive and deliver a baby.)

inhibin B: A hormone produced by the ovaries in females. It is detectable prior to ovulation and reflects the number of follicles remaining in the ovaries (ovarian reserve). In males, it is produced by the testes and is higher in men with normal fertility.

intracytoplasmic sperm injection (ICSI): An IVF procedure in which a single sperm cell is injected directly into the cytoplasm of an egg.

intrauterine insemination (IUI): An ART procedure in which a fine tube is inserted through the cervix into the uterus to directly deposit sperm that have been washed.

in vitro fertilization (IVF): Any medical procedure in which an egg is fertilized by sperm in a test tube. (The key factor: fertilization happens outside the woman's body; *in vitro* means "in glass.")

pelvic inflammatory disease (PID): A disease caused by a sexually transmitted infection that spreads from a woman's vagina to her uterus, fallopian tubes, or ovaries. It is a frequent cause of female infertility.

persistent organic pollutants (POPs): Known as "forever chemicals," these organic compounds remain intact for exceptionally long periods, become widely distributed throughout the environment, accumulate in the fatty tissue of living organisms including humans, and are toxic to humans and wildlife. (This category includes DDT and other pesticides, PCBs, PFASs, and dioxins.)

PFOA, PFOS, and PFASs: Perfluorooctanoic acid (PFOA) and perfluorooctane sulfonate (PFOS) are fluorinated organic compounds that are part of a larger group of compounds known as perfluoroalkyl substances (PFAS). These man-made chemicals are both water- and fat-resistant. They are found in nonstick cookware, stain-resistant carpets, water-resistant clothing, and firefighting foam. Once they're in the environment, they remain there indefinitely.

polybrominated compounds: Polybrominated diphenyl ethers (PBDEs) and polybrominated biphenyls (PBBs) are chemicals that are added to manufactured products (such as furniture, foam padding, wire insulation, rugs, draperies, and upholstery) to reduce the chances that the products will catch on fire. These chemicals can get into the air, water, and soil and build up in certain fish and mammals when they consume contaminated food or water.

polychlorinated biphenyls (PCBs): PCBs are no longer produced in the United States but are still in the environment and can cause health problems. Products made before 1977 that may contain PCBs include old fluorescent lighting fixtures and electrical devices containing PCB capacitors, and old microscope and hydraulic oils. PCBs also are common contaminants in fish.

polycystic ovary syndrome (PCOS): A fairly common hormonal disorder among women of reproductive age. Women with PCOS may have infrequent or prolonged menstrual periods or excess male hormone (androgen) levels and more male-pattern hair distribution.

polyvinyl chloride (PVC): The world's third-most widely produced synthetic plastic polymer. In its rigid form PVC is used in pipes, bottles, food-storage containers, and bank cards. It can be made softer and more flexible (and used, for example, in tubing and plastic wraps) by the addition of plasticizers, the most widely used being phthalates.

progesterone: Known primarily as a female hormone, progesterone is produced in the ovaries, where it plays a key role in the menstrual cycle and prepares the uterus to receive a fertilized egg. In males, the adrenal glands and testes make progesterone, which is needed for testosterone production.

puberty: The period of physical changes through which a child's body matures into an adult body that's capable of sexual reproduction.

rapid-onset gender dysphoria (ROGD): When children suddenly—seemingly out of the blue—decide they identify strongly with the opposite sex.

selective serotonin reuptake inhibitors (SSRIs): Antidepressants that increase levels of serotonin (the "feel-good" hormone) within the brain.

sperm concentration: The number of sperm per milliliter of semen (BTW, this number reflects the number of swimmers in a square area when seen under a microscope—the number should be in the millions).

sperm count (also called total sperm count, or TSC): The total number of sperm in the semen sample a man produces. For math lovers, the equation looks like this: total sperm count = sperm concentration × the volume of the ejaculate sample.

sperm DNA fragmentation index (DFI): The percentage of the sperm that have breaks in their DNA. A high DFI translates into a bad embryo that can fail to implant in the uterus or lead to miscarriage.

sperm morphology: The shape of a man's sperm, including the head, tail, and mid-piece.

sperm motility: Refers to the movement of sperm and their ability to swim. If sperm aren't wriggling vigorously or moving in a straight line, the sperm won't make it to the target.

spontaneous abortion: Also known as miscarriage, *spontaneous abortion* refers to the involuntary loss of the pregnancy anytime between conception and the twentieth week of pregnancy.

stillbirth: Also known as fetal death, which occurs after the twentieth week of pregnancy.

subfertility: A condition resulting in a delay in conceiving. While infertility is the inability to conceive naturally after one year of trying, subfertile couples may conceive naturally but it takes longer than average.

testicular dysgenesis syndrome (TDS): TDS refers to the occurrence at birth of one or more male reproductive conditions: hypospadias, cryptorchidism, poor semen quality, or short AGD; it is associated with an increased risk of testicular cancer and infertility.

transgender: Someone whose gender identity differs from the sex they were assigned at birth.

2,3,7,8-tetrachlorodibenzo-p-dioxin (TCDD): The most toxic form of the chemical dioxin. TCDD accumulates in fat, the placenta, and breast milk. TCDD exposure is linked to low sperm count in men and to endometriosis in women.

varicocele: Enlargement of the veins in the scrotum (like varicose veins); it can reduce a man's fertility.

WHO: World Health Organization, which has been setting the gold standard for semen analysis methods for forty years.

SELECTED BIBLIOGRAPHY

Prologue

Levine, H., N. Jørgensen, A. Martino-Andrade, J. Mendiola, D. Weksler-Derri, I. Mindlis, R. Pinotti, and S. H. Swan. "Temporal trends in sperm count: A systematic review and meta-regression analysis." *Human Reproduction Update* 23(6) (November 2017): 646–59. https://www.ncbi.nlm.nih.gov/pmc/articles/PMC6455044/.

Chapter One:
Reproductive Shock: *Hormonal Havoc in Our Midst*

Carlsen, E., A. Giwercman, N. Keiding, and N. E. Skakkebaek. "Evidence for decreasing quality of semen during past 50 years." *BMJ* 305(6854) (September 12, 1992): 609–13. https://www.ncbi.nlm.nih.gov/pmc/articles/PMC1883354/.

Levine, H., N. Jørgensen, A. Martino-Andrade, J. Mendiola, D. Weksler-Derri, I. Mindlis, R. Pinotti, and S. H. Swan. "Temporal trends in sperm count: A systematic review and meta-regression analysis." *Human Reproduction Update* 23(6) (November 2017): 646–59. https://www.ncbi.nlm.nih.gov/pmc/articles /PMC6455044/.

Swan, S. H., E. P. Elkin, and L. Fenster. "Have sperm densities declined? A reanalysis of global trend data." *Environmental Health Perspectives* 105(11) (1997): 1228–32. https://www.ncbi.nlm.nih.gov/pmc/articles/PMC1470335/.

———. "The question of declining sperm density revisited: An analysis of 101 studies published 1934–1996." *Environmental Health Perspectives* 108(10) (October 2000): 961–66. https://www.ncbi.nlm.nih.gov/pmc/articles/ PMC1240129/.

Chapter Two:
The Diminished Male: *Where Have All the Good Sperm Gone?*

Capogrosso, P., M. Colicchia, E. Ventimiglia, G. Castagna, M. C. Clementi, N. Suardi, F. Castiglione, A. Briganti, F. Cantiello, R. Damiano, F. Montorsi, and A. Salonia. "One patient out of four with newly diagnosed erectile dysfunction is a young man—worrisome picture from the everyday clinical practice." *Journal of Sexual Medicine* 10(7) (July 2013): 1833–41. https://onlinelibrary.wiley.com/doi/full/10.1111/jsm.12179.

Centola, G. M., A. Blanchard, J. Demick, S. Li, and M. L. Eisenberg. "Decline in sperm count and motility in young adult men from 2003 to 2013: Observations from a U.S. sperm bank." *Andrology*, January 20, 2016. https://onlinelibrary.wiley.com/doi/full/10.1111/andr.12149.

Daniels, C. *Exposing Men: The Science and Politics of Male Reproduction.* New York: Oxford University Press, 2006.

Daumler, D., P. Chan, K. C. Lo, J. Takefman, and P. Zelkowitz. "Men's knowledge of their own fertility: A population-based survey examining the awareness of factors that are associated with male infertility." *Human Reproduction* 31(12) (December 2016): 2781–90. https://www.ncbi.nlm.nih.gov/pmc/articles/PMC5193328/.

Dolan, A., T. Lomas, T. Ghobara, and G. Hartshorne. "'It's like taking a bit of masculinity away from you': Towards a theoretical understanding of men's experiences of infertility." *Sociology of Health & Illness* 39(6) (July 2017): 878–92. https://onlinelibrary.wiley.com/doi/full/10.1111/1467-9566.12548.

Fisch, H., G. Hyun, R. Golden, R. W. Hensle, C. A. Olsson, and G. L. Liberson. "The influence of paternal age on down syndrome." *Journal of Urology* 169(6) (June 2003): 2275–78. https://www.ncbi.nlm.nih.gov/pubmed/12771769.

Goisis, A., H. Remes, P. Martikainen, R. Klemetti, and M. Myrskylä. "Medically assisted reproduction and birth outcomes: A within-family analysis using Finnish population registers." *Lancet* 393(10177) (March 23, 2019): 1225–32. https://www.ncbi.nlm.nih.gov/pubmed/30655015.

Grand View Research. "Sperm bank market size analysis report by service type (sperm storage, semen analysis, genetic consultation), by donor type (known, anonymous), by end use, and segment forecasts, 2019–2025." May 2019. https://www.grandviewresearch.com/industry-analysis/sperm-bank-market.

———. "Sperm bank market worth $5.45 billion by 2025." May 2019. https://www.grandviewresearch.com/press-release/global-sperm-bank-market.

Guzick, D. S., J. W. Overstreet, P. Factor-Litvak, C. K. Brazil, S. T. Nakajima, C. Coutifaris, S. A. Carson et al. "Sperm morphology, motility, and concentration in fertile and infertile men." *New England Journal of Medicine* 345(19) (November 8, 2001): 1388–93.

Hsieh, F-I., T-S. Hwang, Y-C. Hsieh, H-C. Lo, C-T. Su, H-S. Hsu, H-Y. Chiou, and C-J. Chen. "Risk of erectile dysfunction induced by arsenic exposure through well water consumption in Taiwan." *Environmental Health Perspectives* 116(4) (April 2008): 532–36. https://www.ncbi.nlm.nih.gov/pmc/articles /PMC2291004/.

Huang, C., B. Li, K. Xu, D. Liu, J. Hu, Y. Yang, H. C. Nie, L. Fan, and W. Zhu. "Decline in semen quality among 30,636 young Chinese men from 2001 to 2015." *Fertility and Sterility* 107(1) (January 2017): 83–88. https://www.fertstert.org/article /S0015-0282(16)62866-2/pdf.

Inhorn, M. C., and P. Patrizio. "Infertility around the globe: New thinking on gender, reproductive technologies and global movements in the 21st century." *Human Reproduction Update* 21(4) (July/August 2015): 411–26. https://academic.oup .com/humupd/article/21/4/411/683746.

Kleinhaus, K., M. Perrin, Y. Friedlander, O. Paltiel, D. Malaspina, and S. Harlap. "Paternal age and spontaneous abortion." *Obstetrics and Gynecology* 108(2) (August 2006): 369–77. https://www.ncbi.nlm.nih.gov/pubmed/16880308.

Marin Fertility Center. "Infertility basics." http://marinfertilitycenter.com/new -getting-started/infertility-basics/.

May, G. "Erectile dysfunction is on the rise among young men and here's why." *Marie Claire*, March 13, 2018. https://www.marieclaire.co.uk/life/sex-and -relationships/erectile-dysfunction-579283.

Oliva, A., A. Giami, and L. Multigner. "Environmental agents and erectile dysfunction: A study in a consulting population." *Journal of Andrology* 23(4) (July–August 2002): 546–50. https://www.ncbi.nlm.nih.gov/pubmed/12065462.

Planned Parenthood. "When do boys start producing sperm?" October 5, 2010. https://www.plannedparenthood.org/learn/teens/ask-experts/when-do-boys -start-producing-sperm.

Rais, A., S. Zarka, E. Derazne, D. Tzur, R. Calderon-Margalit, N. Davidovitch, A. Afek, R. Carel, and H. Levine. "Varicocoele among 1,300,000 Israeli adolescent males: Time trends and association with body mass index." *Andrology* 1(5) (September 2013): 663–69. https://onlinelibrary.wiley.com/doi/ full/10.1111/j.2047-2927.2013.00113.x.

Richard, J., I. Badillo-Amberg, and P. Zelkowitz. "'So much of this story could be me': Men's use of support in online infertility discussion boards." *American Journal of Men's Health* 11(3) (2017): 663–73. https://journals.sagepub.com/doi /pdf/10.1177/1557988316671460.

Slama, R., J. Bouyer, G. Windham, L. Fenster, A. Werwatz, and S. H. Swan. "Influence of paternal age on the risk of spontaneous abortion." *American Journal of Epidemiology* 161(9) (May 1, 2005): 816–23. https://www.ncbi.nlm.nih.gov/pubmed/15840613.

Smith, J. F., T. J. Walsh, A. W. Shindel, P. J. Turek, H. Wing, L. Pasch, P. P. Katz, and the Infertility Outcomes Project Group. "Sexual, marital, and social impact of a man's perceived infertility diagnosis." *Journal of Sexual Medicine* 6(9) (September 2009): 2505–15. https://www.ncbi.nlm.nih.gov/pmc/articles/PMC2888139/.

Tiegs, A., J. Landis, N. Garrido, R. Scott, and J. Hotaling. "Total motile sperm count trend over time: Evaluation of semen analyses from 119,972 subfertile men from 2002 to 2017." *Urology* 132 (October 2019): 109–16. https://www.ncbi .nlm.nih.gov/pubmed/31326545.

Chapter Three:
It Takes Two to Tango: *Her Side of the Story*

Aksglaede, L., K. Sørensen, J. H. Petersen, N. E. Skakkebaek, and A. Juul. "Recent decline in age at breast development: The Copenhagen Puberty Study." *Pediatrics* 123(5) (May 2009): e932–e939. https://www.ncbi.nlm.nih.gov/pubmed/19403485.

American College of Obstetricians and Gynecologists. "Early pregnancy loss." Practice Bulletin, November 2018. https://www.acog.org/Clinical-Guidance-and -Publications/Practice-Bulletins/Committee-on-Practice-Bulletins-Gynecology /Early-Pregnancy-Loss.

———. "Female age-related fertility decline." Committee Opinion, March 2014. https://www.acog.org/Clinical-Guidance-and-Publications/Committee-Opinions /Committee-on-Gynecologic-Practice/Female-Age-Related-Fertility-Decline.

American Psychological Association. "The risks of earlier puberty." March 2016. https://www.apa.org/monitor/2016/03/puberty.

"Ava International Fertility & TTC 2017 Report." Press release, September 13, 2017. https://3xwa2438796x1hj4o4m8vrk1-wpengine.netdna-ssl.com/wp-content /uploads/2017/09/Ava-Fertility-Survey-Press-Release.pdf.

Balasch, J. "Ageing and infertility: An overview." *Gynecological Endocrinology* 26(12) (December 2010): 855–60. https://www.ncbi.nlm.nih.gov/pubmed/20642380.

Bjelland, E. K., S. Hofvind, L. Byberg, and A. Eskild. "The relation of age at menarche with age at natural menopause: A population study of 336,788 women in Nor- way." *Human Reproduction* 33(6) (June 1, 2018): 1149–57. https://www.ncbi.nlm .nih.gov/pmc/articles/PMC5972645/.

BMJ Best Practice. "Precocious puberty." Last reviewed February 2020. https://best practice.bmj.com/topics/en-us/1127.

Bretherick, K. L., N. Fairbrother, L. Avila, S. H. Harbord, and W. P. Robinson. "Fer- tility and aging: Do reproductive-aged Canadian women know what they need to know?" *Fertility and Sterility* 93(7) (May 2010): 2162–68. https://www.ncbi .nlm.nih.gov/pubmed/19296943.

Brix, N., A. Ernst, L. L. B. Lauridsen, E. Parner, H. Støvring, J. Olsen, T. B. Henrik- sen, and C. H. Ramlau-Hansen. "Timing of puberty in boys and girls: A population-based study." *Paediatric and Perinatal Epidemiology* 33(1) (January 2019): 70–78. https://www.ncbi.nlm.nih.gov/pmc/articles/PMC6378593/.

Cedars, M. I., S. E. Taymans, L. V. DePaolo, L. Warner, S. B. Moss, and M. Eisenberg. "The sixth vital sign: What reproduction tells us about overall health. Proceedings from a NICHD/CDC workshop." *Human Reproduction Open*, 2017, 1–8. https://urology.stanford.edu/content/dam/sm/urology/JJimages/publications /The-sixth-vital-sign-what-reproduction-tells-us-about-overall-health-Proceed ings-from-a-NICHD-CDC-workshop.pdf.

Devine, K., S. L. Mumford, M. Wu, A. H. DeCherney, M. J. Hill, and A. Propst. "Diminished ovarian reserve (DOR) in the US ART population: Diagnostic trends among 181,536 cycles from the Society for Assisted Reproductive Technology Clinic Outcomes Reporting System (SART CORS)." *Fertility and Sterility* 104(3) (September 2015): 612–19. https://www.ncbi.nlm.nih.gov/pmc /articles/PMC4560955/.

Gleicher, N., V. A. Kushnir, A. Weghofer, and D. H. Barad. "The 'graying' of infertility services: An impending revolution nobody is ready for." *Reproductive Biology and Endocrinology* 12 (2014): 63. https://www.ncbi.nlm.nih.gov/pmc/articles /PMC4105876/.

Gossett, D. R., S. Nayak, S. Bhatt, and S. C. Bailey. "What do healthy women know about the consequences of delayed childbearing?" *Journal of Health Communication* 18(Suppl 1) (December 2013): 118–28. https://www.ncbi.nlm.nih .gov/pmc/articles/PMC3814907/.

Grand View Research. "Assisted reproductive technology (ART) market size, share & trends analysis report by type (IVF, others), by end use (hospitals, fertility clinics), by procedures and segment forecasts, 2018–2025." May 2018. https:// www.grandviewresearch.com/industry-analysis/assisted-reproductive-tech nology-market.

Harrington, R. "Elective human egg freezing on the rise." *Scientific American*, February 18, 2015. https://www.scientificamerican.com/article/elective-human -egg-freezing-on-the-rise/.

Hayden, E. C. "Cursed Royal Blood: Was Henry VIII to blame for his wives' many miscarriages?" *Slate*, May 15, 2013. https://slate.com/technology/2013/05/henry -viii-wives-and-children-were-kell-proteins-to-blame-for-many-miscarriages .html.

Herman-Giddens, M. E., E. J. Slora, R. C. Wasserman, C. J. Bourdony, M. V. Bhapkar, G. G. Koch, and C. M. Hasemeier. "Secondary sexual characteristics and menses in young girls seen in office practice: A study from the pediatric research in office settings network." *Pediatrics* 99(4) (April 1997): 505–12. https://www.ncbi.nlm.nih.gov/pubmed/9093289.

Hosokawa, M., S. Imazeki, H. Mizunuma, T. Kubota, and K. Hayashi. "Secular trends in age at menarche and time to establish regular menstrual cycling in Japanese women born between 1930 and 1985." *BMC Womens Health* 12(19) (2012). https://www.ncbi.nlm.nih.gov/pmc/articles/PMC3434095/.

Hunter, A., L. Tussis, and A. MacBeth. "The presence of anxiety, depression and stress in women and their partners during pregnancies following perinatal loss: A meta-analysis." *Journal of Affective Disorders* 223 (December 2017): 153–64. https://www.ncbi.nlm.nih.gov/pubmed/28755623.

InterLACE Study Team. "Variations in reproductive events across life: A pooled analysis of data from 505,147 women across 10 countries." *Human Reproduction* 34(5) (March 2019): 881–93. https://www.ncbi.nlm.nih.gov/pubmed/30835788.

Jayasena, C. N., U. K. Radia, M. Figueiredo, L. F. Revill, A. Dimakopoulou, M. Osagie, W. Vessey, L. Regan, R. Rai, and W. S. Dhillo. "Reduced testicular steroidogenesis and increased semen oxidative stress in male partners as novel markers of recurrent miscarriage." *Clinical Chemistry* 65(1) (2019): 161–69. https://www.ncbi.nlm.nih.gov/pubmed/30602480.

Jensen, M. B., L. Priskorn, T. K. Jensen, A. Juul, and N. E. Skakkebaek. "Temporal trends in fertility rates: A nationwide registry based study from 1901 to 2014." *PLoS One* 10(12) (2015): e0143722. https://www.ncbi.nlm.nih.gov/pmc/articles/PMC4668020/.

Kinsey, C. B., K. Baptiste-Roberts, J. Zhu, and K. H. Kjerulff. "Effect of previous miscarriage on depressive symptoms during subsequent pregnancy and postpartum in the first baby study." *Maternal and Child Health Journal* 19(2) (February 2015): 391–400. https://www.ncbi.nlm.nih.gov/pmc/articles/PMC4256135/.

Kolte, A. M., L. R. Olsen, E. M. Mikkelsen, O. B. Christiansen, and H. S. Nielsen. "Depression and emotional stress is highly prevalent among women with recurrent pregnancy loss." *Human Reproduction* 30(4) (April 2015): 777–82. https://www.ncbi.nlm.nih.gov/pmc/articles/PMC4359400/.

Kudesia, R., E. Chernyak, and B. McAvey. "Low fertility awareness in United States reproductive-aged women and medical trainees: Creation and validation of the Fertility & Infertility Treatment Knowledge Score (FIT-KS)." *Fertility and Sterility* 108(4) (October 2017): 711–17. https://www.ncbi.nlm.nih.gov/pubmed/28911930.

Lundsberg, L. S., L. Pal, A. M. Gariepy, X. Xu, M. C. Chu, and J. L. Illuzzi. "Knowledge, attitudes, and practices regarding conception and fertility: A population-based survey among reproductive-age United States women." *Fertility and Sterility* 101(3) (March 2014): 767–74. https://www.ncbi.nlm.nih.gov/pubmed/24484995.

Matthews, T. J., and B. E. Hamilton. "Total fertility rates by state and race and Hispanic origin: United States, 2017." *National Vital Statistics Reports* 68(1) (January 2019): 1–11. https://www.ncbi.nlm.nih.gov/pubmed/30707671.

Menasha, J., B. Levy, K. Hirschhorn, and N. B. Kardon. "Incidence and spectrum of chromosome abnormalities in spontaneous abortions: New insights from a 12-year study." *Genetics in Medicine* 7(4) (April 2005): 251–63. https://www.ncbi.nlm.nih.gov/pubmed/15834243.

Mendle, J., E. Turkheimer, and R. E. Emery. "Detrimental psychological outcomes associated with early pubertal timing in adolescent girls." *Developmental Review* 27(2) (June 2007): 151–71. https://www.ncbi.nlm.nih.gov/pmc /articles/PMC2927128/.

Obama, M. *Becoming.* New York: Crown, 2018.

O'Connor, K. A., D. J. Holman, and J. W. Wood. "Declining fecundity and ovarian ageing in natural fertility populations." *Maturitas* 30 (2) (October 1998): 127–36. https://www.ncbi.nlm.nih.gov/pubmed/9871907.

Paris, K., and A. Aris. "Endometriosis-associated infertility: A decade's trend study of women from the Estrie region of Quebec, Canada." *Gynecological Endocrinology* 26(11) (November 2010): 838–42. https://www.ncbi.nlm.nih.gov /pubmed/20486880.

Perkins, K. M., S. L. Boulet, D. J. Jamieson, D. M. Kissin; National Assisted Reproductive Technology Surveillance System Group. "Trends and outcomes of gestational surrogacy in the United States." *Fertility and Sterility* 106(2) (August 2016): 435–42. https://www.ncbi.nlm.nih.gov/pubmed/27087401.

Practice Committee of the American Society for Reproductive Medicine. "Testing and interpreting measures of ovarian reserve: A committee opinion." *Fertility and Sterility* 103(3) (March 2015): e9–e17. https://www.fertstert.org/article /S0015-0282(14)02518-7/pdf.

Pylyp, L. Y., L. O. Spynenko, N. V. Verhoglyad, A. O. Mishenko, D. O. Mykytenko, and V. D. Zukin. "Chromosomal abnormalities in products of conception of first-trimester miscarriages detected by conventional cytogenetic analysis: A review of 1,000 cases." *Journal of Assisted Reproduction and Genetics* 35(2) (February 2018): 265–71. https://www.ncbi.nlm.nih.gov/pmc/articles/PMC5845039/.

Roepke, E. R., L. Matthiesen, R. Rylance, and O. B. Christiansen. "Is the incidence of recurrent pregnancy loss increasing? A retrospective register-based study in Sweden." *Acta Obstetricia et Gynecolgica Scandinavica* 96(11) (November 2017): 1365–72. https://obgyn.onlinelibrary.wiley.com/doi/full/10.1111/aogs.13210.

Rossen, L. M., K. A. Ahrens, and A. M. Branum. "Trends in risk of pregnancy loss among US women, 1990–2011." *Paediatric and Perinatal Epidemiology* 32 (1) (January 2018): 19–29. https://www.ncbi.nlm.nih.gov/pmc/articles /PMC5771868/.

Swan, S. H., I. Hertz-Picciotto, A. Chandra, and E. H. Stephen. "Reasons for infecundity." *Family Planning Perspectives* 31(3) (May–June 1999): 156–57. https://www.jstor.org/stable/2991707?seq=1.

Swift, B. E., and K. E. Liu. "The effect of age, ethnicity, and level of education on fertility awareness and duration of infertility." *Journal of Obstetrics and Gynaecology Canada* 36(11) (November 2014): 990–96. https://www.ncbi.nlm.nih.gov /pubmed/25574676.

Tavoli, Z., M. Mohammadi, A. Tavoli, A. Moini, M. Effatpanah, L. Khedmat, and A. Montazeri. "Quality of life and psychological distress in women with recurrent miscarriage: A comparative study." *Health and Quality of Life Outcomes*, July 2018. https://www.ncbi.nlm.nih.gov/pmc/articles/PMC6064101/.

Thomas, H. N., M. Hamm, R. Hess, S. Borreoro, and R. C. Thurston. "'I want to feel like I used to feel': A qualitative study of causes of low libido in postmenopausal women." *Menopause* 27(3) (March 2020): 289–94. https://www.ncbi.nlm.nih.gov/pubmed/31834161.

WebMD. "What is a normal period?" https://www.webmd.com/women/normal-period.

Wilcox, A. J., C. R. Weinberg, J. F. O'Connor, D. D. Baird, J. P. Schlatterer, R. E. Canfield, E. G. Armstrong, and B. C. Nisula. "Incidence of early loss of pregnancy." *New England Journal of Medicine* 319(4) (July 28, 1988): 189–94. https://www.ncbi.nlm.nih.gov/pubmed/3393170.

World Bank. "Fertility rate, total (births per woman)—United States." https://data.worldbank.org/indicator/SP.DYN.TFRT.IN?locations=US.

Worsley, R., R. J. Bell, P. Gartoulla, and S. R. Davis. "Prevalence and predictors of low sexual desire, sexually related personal distress, and hypoactive sexual desire dysfunction in a community-based sample of midlife women." *Journal of Sexual Medicine* 14(5) (May 2017): 675–86. https://www.jsm.jsexmed.org/article/S1743-6095(17)30418-6/fulltext.

Yu, L., B. Peterson, M. C. Inhorn, J. K. Boehm, and P. Patrizio. "Knowledge, attitudes, and intentions toward fertility awareness and oocyte cryopreservation among obstetrics and gynecology resident physicians." *Human Reproduction* 31(2) (February 2016): 402–11. https://www.ncbi.nlm.nih.gov/pubmed/26677956.

Chapter Four:
Gender Fluidity: *Beyond Male and Female*

Airton, L. *Gender: Your Guide.* Avon, MA: Adams Media, 2018.

American Psychological Association. "Answers to your questions about individuals with intersex conditions." https://www.apa.org/topics/lgbt/intersex.pdf.

Bejerot, S., M. B. Humble, and A. Gardner. "Endocrine disruptors, the increase of autism spectrum disorder and its comorbidity with gender identity disorder—a hypothetical association." *International Journal of Andrology* 34(5 pt.2) (October 2011): e350. https://onlinelibrary.wiley.com/doi/full/10.1111/j.1365-2605.2011.01149.x.

Berenbaum, S. A. "Beyond pink and blue: The complexity of early androgen effects on gender development." *Child Development Perspectives* 12(1) (March 2018): 58–64. https://www.ncbi.nlm.nih.gov/pmc/articles/PMC5935256/.

Berenbaum, S. A., and E. Snyder. "Early hormonal influences on childhood sex-typed activity and playmate preferences: Implications for the development of sexual orientation." *Developmental Psychology* 3(1) (1995): 31–42. https://psycnet.apa.org/doiLanding?doi=10.1037%2F0012-1649.31.1.31.

Children's National. "Pediatric differences in sex development." https://childrens national.org/visit/conditions-and-treatments/diabetes-hormonal-disorders /differences-in-sex-development.

Dastagir, A. E. "'Born this way'? It's way more complicated than that." *USA Today*, June 15, 2017. https://www.usatoday.com/story/news/2017/06/16/born -way-many-lgbt-community-its-way-more-complex/395035001/.

Ehrensaft, D. *Gender Born, Gender Made: Raising Healthy Gender-Nonconforming Children.* New York: Experiment, 2011.

Gaspari, L., F. Paris, C. Jandel, N. Kalfa, M. Orsini, J. P. Daurès, and C. Sultan. "Prenatal environmental risk factors for genital malformations in a population of 1,442 French male newborns: A nested case-control study." *Human Reproduction* 26(11) (November 2011): 3155–62. https://www.ncbi.nlm.nih.gov/pubmed/21868402.

Glidden, D., W. P. Bouman, B. A. Jones, and J. Arcelus. "Gender dysphoria and autism spectrum disorder: A systematic review of the literature." *Sexual Medicine Reviews* 4(1) (January 2016): 3–14. https://www.ncbi.nlm.nih.gov/pubmed/27872002.

Hadhazy, A. "What makes Michael Phelps so good?" *Scientific American*, August 18, 2008. https://www.scientificamerican.com/article/what-makes-michael-phelps-so-good1/.

Hedaya, R. J. "The dissolution of gender: The role of hormones." *Psychology Today*, February 13, 2019. https://www.psychologytoday.com/us/blog/health-matters /201902/the-dissolution-gender.

Intersex Society of North America. "How common is intersex?" http://www.isna .org/faq/frequency.

———. "What is intersex?" http://www.isna.org/faq/what_is_intersex.

Ives, M. "Sprinter Dutee Chand Becomes India's First Openly Gay Athlete." *New York Times*, May 20, 2019. https://www.nytimes.com/2019/05/20/world/asia /india-dutee-chand-gay.html.

Katwala, A. "The controversial science behind the Caster Semenya verdict." *Wired*, May 1, 2019. https://www.wired.co.uk/article/caster-semenya-testosterone -ruling-gender-science-analysis.

Kazemian, L. "Desistance." *Oxford Bibliographies*. Last reviewed April 21, 2017. https://www.oxfordbibliographies.com/view/document/obo-9780195396607 /obo-9780195396607-0056.xml.

Keating, S. "Gender dysphoria isn't a 'social contagion,' according to a new study." *BuzzFeed News*, April 22, 2019. https://www.buzzfeednews.com/article /shannonkeating/rapid-onset-gender-dysphoria-flawed-methods-transgender.

Lehrman, S. "When a person is neither XX nor XY: A Q & A with geneticist Eric Vilain." *Scientific American*, May 30, 2007. https://www.scientificamerican.com /article/q-a-mixed-sex-biology/.

Littman, L. "Parent reports of adolescents and young adults perceived to show signs of a rapid onset of gender dysphoria." *PLoS One* 13(8) (August 16, 2018): e0202330. https://journals.plos.org/plosone/article?id=10.1371/journal.pone.0202330.

Magliozzi, D., A. Saperstein, and L. Westbrook. "Scaling up: Representing gender diversity in survey research." *Socius: Sociological Research for a Dynamic World*, August 19, 2016. https://journals.sagepub.com/doi/10.1177/2378023116664352.

Mukherjee, S. *The Gene: An Intimate History*. New York: Scribner, 2016.

Nakagami, A., T. Negishi, K. Kawasaki, N. Imai, Y. Nishida, T. Ihara, Y. Kuroda, Y. Yoshikawa, and T. Koyama. "Alterations in male infant behaviors towards its mother by prenatal exposure to bisphenol A in cynomolgus monkeys (*Macaca fascicularis*) during early suckling period." *Psychoneuroendocrinology* 34(8) (2009): 1189–97. https://www.ncbi.nlm.nih.gov/pubmed/19345509.

Newhook, J. T., J. Pyne, K. Winters, S. Feder, C. Holmes, J. Tosh, M-L Sinnott, A. Jamieson, and S. Pickett. "A critical commentary on follow-up studies and 'desistance' theories about transgender and gender-nonconforming children." *International Journal of Transgenderism* 19(2) (2018): 212–24. https://www.tandfonline.com/doi/abs/10.1080/15532739.2018.1456390.

Newport, F. "In U.S., estimate of LGBT population rises to 4.5%." Gallup, May 22, 2018. https://news.gallup.com/poll/234863/estimate-lgbt-population-rises.aspx.

Padawer, R. "The humiliating practice of sex-testing female athletes." *New York Times*, June 28, 2016. https://www.nytimes.com/2016/07/03/magazine/the-humiliating-practice-of-sex-testing-female-athletes.html?_r=0.

Pasterski, V. L., M. E. Geffner, C. Brain, P. Hindmarsh, C. Brook, and M. Hines. "Prenatal hormones and postnatal socialization by parents as determinants of male-typical toy play in girls with congenital adrenal hyperplasia." *Child Development* 76(1) (January–February 2005): 264–78. https://www.ncbi.nlm.nih.gov/pubmed/15693771.

Restar, A. J. "Methodological critique of Littman's (2018) parental-respondents accounts of 'rapid-onset gender dysphoria.'" *Archives of Sexual Behavior* 49 (2020): 61–66. https://link.springer.com/article/10.1007/s10508-019-1453-2.

Rich, A. L., L. M. Phipps, S. Tiwari, H. Rudraraju, and P. O. Dokpesi. "The increasing prevalence in intersex variation from toxicological dysregulation in fetal reproductive tissue differentiation and development by endocrine-disrupting chemicals." *Environmental Health Insights* 10 (2016): 163–71. https://www.ncbi.nlm.nih.gov/pmc/articles/PMC5017538/.

Saguy, A. C., J. A. Williams, R. Dembroff, and D. Wodak. "We should all use they/them pronouns . . . eventually." *Scientific American*, May 30, 2019. https://blogs.scientificamerican.com/voices/we-should-all-use-they-them-pronouns-eventually/.

Saperstein, A. "State of the Union 2018: Gender identification." *Stanford Center on Poverty and Inequality*. https://inequality.stanford.edu/sites/default/files/Pathways_SOTU_2018_gender-ID.pdf.

"Swiss court blocks Semenya from 800 at worlds." Associated Press, July 30, 2019. https://www.espn.com/olympics/trackandfield/story/_/id/27288611/swiss-court-blocks-semenya-800-worlds.

Tobia, J. *Sissy: A Coming-of-Gender Story*. New York: Putnam, 2019.

Vandenbergh, J. G., and C. L. Huggett. "The anogenital distance index, a predictor of the intrauterine position effects on reproduction in female house mice." *Laboratory Animal Science* 45(5) (October 1995): 567–73. https://www.ncbi.nlm.nih.gov/pubmed/8569159.

"What is congenital adrenal hyperplasia?" You and Your Hormones. https://www.yourhormones.info/endocrine-conditions/congenital-adrenal-hyperplasia/.

Chapter Five:
Windows of Vulnerability: *Timing Is Everything*

Axelsson J., S. Sabra, L. Rylander, A. Rignell-Hydbom, C. H. Lindh, and A. Giwercman. "Association between paternal smoking at the time of pregnancy and the semen quality in sons." *PLoS ONE* 13(11) (November 21, 2018): e0207221. https://www.ncbi.nlm.nih.gov/pmc/articles/PMC6248964/.

Bell, M. R., L. M. Thompson, K. Rodriguez, and A. C. Gore. "Two-hit exposure to polychlorinated biphenyls at gestational and juvenile life stages: 1. Sexually dimorphic effects on social and anxiety-like behaviors." *Hormones and Behavior* 78 (February 2016): 168–77. https://www.ncbi.nlm.nih.gov/pubmed/26592453.

Binder, A. M., C. Corvalan, A. Pereira, A. M. Calafat, X. Ye, J. Shepherd, and K. B. Michels. "Prepubertal and pubertal endocrine-disrupting chemical exposure and breast density among Chilean adolescents." *Cancer Epidemiology, Biomarkers & Prevention* 27(12) (December 2018): 1491–99. https://www.ncbi.nlm.nih.gov/pmc/articles/PMC6541222/.

Bräuner, E. V., D. A. Doherty, J. E. Dickinson, D. J. Handelsman, M. Hickey, N. E. Skakkebaek, A. Juul, and R. Hart. "The association between in-utero exposure to stressful life events during pregnancy and male reproductive function in a cohort of 20-year-old offspring: The Raine Study." *Human Reproduction* 34(7) (July 8, 2019): 1345–55. https://www.ncbi.nlm.nih.gov/pubmed/31143949.

Dees, W. L., J. K. Hiney, and V. K. Srivastava. "Alcohol and puberty." *Alcohol Research* 38(2) (2017): 277–82. https://www.ncbi.nlm.nih.gov/pmc/articles/PMC5513690/.

Dranow, D. B., R. P. Tucker, and B. W. Draper. "Germ cells are required to maintain a stable sexual phenotype in adult zebrafish." *Developmental Biology* 376:43–50. http://thenode.biologists.com/sex-reversal-in-adult-fish/research/.

Durmaz, E., E. N. Ozmert, P. Erkekoglu, B. Giray, O. Derman, F. Hincal, and K. Yurdakök. "Plasma phthalate levels in pubertal gynecomastia." *Pediatrics* 125(1) (January 2010): e122–e129. https://www.ncbi.nlm.nih.gov/pubmed/20008419.

Edwards, A., A. Megens, M. Peek, and E. M. Wallace. "Sexual origins of placental dysfunction." *Lancet* 355(9199) (January 15, 2000): 203–4. www.thelancet.com/journals/lancet/article/PIIS0140-6736(99)05061-8/fulltext.

Eriksson, J. G., E. Kajantie, C. Osmond, K. Thornburg, and D. J. P. Barker. "Boys live dangerously in the womb." *American Journal of Human Biology* 22(3) (2010): 330–35. https://www.ncbi.nlm.nih.gov/pmc/articles/PMC3923652 /pdf/nihms240904.pdf.

"5 crazy things doctors used to tell pregnant women." Kodiak Birth and Wellness, November 9, 2016. http://birthgoals.com/blog/2016/7/25/5-surprising -things-doctors-used-to-tell-pregnant-women.

Gray, L. E, Jr., V. S. Wilson, T. E. Stoker, C. R. Lambright, J. R. Furr, N. C. Noriega, P. C. Hartig et al. "Environmental androgens and antiandrogens: An expanding chemical universe." EPA Home, Science Inventory, 2004, 313–45. https://cfpub .epa.gov/si/si_public_record_report.cfm?dirEntryId=104084&Lab=NHEERL.

Grech, V. "Terrorist attacks and the male-to-female ratio at birth: The Troubles in Northern Ireland, the Rodney King riots, and the Breivik and Sandy Hook shootings." *Early Human Development* 91(12) (December 2015): 837–40. www.ncbi.nlm.nih.gov/pubmed/26525896.

Hill, M. A. "Timeline human development." Embryology. https://embryology.med .unsw.edu.au/embryology/index.php/Timeline_human_development.

Lund, L., M. C. Engebjerg, L. Pedersen, V. Ehrenstein, M. Nørgaard, and H. T. Sørensen. "Prevalence of hypospadias in Danish boys: A longitudinal study, 1977–2005." *European Urology* 55(5) (May 2009): 1022–26. https://www.ncbi .nlm.nih.gov/pubmed/19155122.

MacLeod, D. J., R. M. Sharpe, M. Welsh, M. Fisken, H. M. Scott, G .R. Hutchison, A. J. Drake, and S. van den Driesche. "Androgen action in the masculinization programming window and development of male reproductive organs." *International Journal of Andrology* 33(2) (April 2010): 279–87. https://www.ncbi.nlm .nih.gov/pubmed/20002220.

Martino-Andrade, A. J., F. Liu, S. Sathyanarayana, E. S. Barrett, J. B. Redmon, R. H. Nguyen, H. Levine, S. H. Swan, and the TIDES Study Team. "Timing of prenatal phthalate exposure in relation to genital endpoints in male newborns." *Andrology* 4(4) (July 2016): 585–93. https://www.ncbi.nlm.nih.gov /pubmed/27062102.

Masukume, G., S. M. O'Neill, A. S. Khashan, L. C. Kenny, and V. Grech. "The terrorist attacks and the human live birth sex ratio: A systematic review and meta-analysis." *Acta Medica* 60(2) (2017): 59–65. https://actamedica.lfhk.cuni .cz/media/pdf/am_2017060020059.pdf.

Mínguez-Alarcón, L., I. Souter, Y-H. Chiu, P. L. Williams, J. B. Ford, A. Ye, A. M. Calafat, and R. Hauser. "Urinary concentrations of cyclohexane-1,2- dicarboxylic acid monohydroxy isononyl ester, a metabolite of the non-phthalate plasticizer di(isononyl)cyclohexane-1,2-dicarboxylate (DINCH), and markers of ovarian response among women attending a fertility center." *Environmental*

Research 151 (November 2016): 595–600. https://www.ncbi.nlm.nih.gov/pmc /articles/PMC5071161/.

National Cancer Institute. "Diethylstilbestrol (DES) and cancer." Reviewed October 5, 2011. https://www.cancer.gov/about-cancer/causes-prevention/risk /hormones/des-fact-sheet.

Nordenvall, A. S., L. Frisén, A. Nordenström, P. Lichtenstein, and A. Nordenskjöld. "Population based nationwide study of hypospadias in Sweden, 1973 to 2009: Incidence and risk factors." *Journal of Urology* 191(3) (March 2014): 783–89. https://www.ncbi.nlm.nih.gov/pubmed/24096117.

Olson, E. R. "Why are 250 million sperm cells released during sex?" LiveScience, January 24, 2013. https://www.livescience.com/32437-why-are-250-million -sperm-cells-released-during-sex.html.

Pasterski, V., C. L. Acerini, D. B. Dunger, K. K. Ong, I. A. Hughes, A. Thankamony, and M. Hines. "Postnatal penile growth concurrent with mini-puberty predicts later sex-typed play behavior: Evidence for neurobehavioral effects of the post-natal androgen surge in typically developing boys." *Hormones and Behavior* 69 (March 2015): 98–105. https://www.ncbi.nlm.nih.gov/pubmed/25597916.

Pennisi, E. "Why women's bodies abort males during tough times." *Science*, December 11, 2014. https://www.sciencemag.org/news/2014/12/why-women-s -bodies-abort-males-during-tough-times.

Roy, P., A. Kumar, I. R. Kaur, and M. M. Faridi. "Gender differences in outcomes of low birth weight and preterm neonates: The male disadvantage." *Journal of Tropical Pediatrics* 60(6) (December 2014): 480–81. https://www.ncbi.nlm.nih .gov/pubmed/25096219.

SexInfoOnline. "Sex determination and differentiation." Last updated November 3, 2016. http://www.soc.ucsb.edu/sexinfo/article/sex-determination-and -differentiation.

Skakkebaek, N. E., E. Rajpert–De Meyts, G. M. Buck Louis, J. Toppari, A. M. Andersson, M. L. Eisenberg, T. K. Jensen. "Male reproductive disorders and fertility trends: Influences of environment and genetic susceptibility." *Physiological Reviews* 96(1) (January 2016): 55–97. https://www.ncbi.nlm.nih.gov/pmc /articles/PMC4698396/.

Swan, S. H., K. M. Main, F. Liu, S. L. Stewart, R. L. Kruse, A. M. Calafat, C. S. Mao et al. "Decrease in anogenital distance among male infants with prenatal phthalate exposure." *Environmental Health Perspectives* 113(8) (2005): 1056–61. https://www.ncbi.nlm.nih.gov/pmc/articles/PMC1280349/.

Wu, Y., G. Zhong, S. Chen, C. Zheng, D. Liao, and M. Xie. "Polycystic ovary syndrome is associated with anogenital distance, a marker of prenatal androgen exposure." *Human Reproduction* 32(4) (April 1, 2017): 937–43. https://www .ncbi.nlm.nih.gov/pubmed/28333243.

Chapter Six:
Up Close and Personal: *Lifestyle Habits That Can Sabotage Fertility*

Afeiche, M., A. J. Gaskins, P. L. Williams, T. L. Toth, D. L. Wright, C. Tanrikut, R. Hauser, and J. E. Chavarro. "Processed meat intake is unfavorably and fish intake favorably associated with semen quality indicators among men attending a fertility clinic." *Journal of Nutrition* 144(7) (July 2014): 1091–98. https://www.ncbi .nlm.nih.gov/pmc/articles/PMC4056648/.

Afeiche, M. C., P. L. Williams, A. J. Gaskins, J. Mendiola, N. Jørgensen, S. H. Swan, and J. E. Chavarro. "Meat intake and reproductive parameters among young men." *Epidemiology* 25(3) (May 2014): 323–30. https://www.ncbi.nlm.nih.gov/pmc/articles /PMC4180710/.

Afeiche, M., P. L. Williams, J. Mendiola, A. J. Gaskins, N. Jørgensen, S. H. Swan, and J. E. Chavarro. "Dairy food intake in relation to semen quality and reproductive hormone levels among physically active young men." *Human Reproduction* 28(8) (August 2013): 2265–75. https://www.ncbi.nlm.nih.gov/pmc/articles /PMC3712661/.

American Academy of Orthopaedic Surgeons. "Female athlete triad: Problems caused by extreme exercise and dieting." Last reviewed June 2016. https://ortho info.aaos.org/en/diseases--conditions/female-athlete-triad-problems-caused -by-extreme-exercise-and-dieting/.

American Society for Reproductive Medicine. "Third-party reproduction: A guide for patients." Revised 2017. https://www.reproductivefacts.org/globalassets /rf/news-and-publications/bookletsfact-sheets/english-fact-sheets-and-info -booklets/third-party_reproduction_booklet_web.pdf.

Bae, J., S. Park, and J-W. Kwon. "Factors associated with menstrual cycle irregularity and menopause." *BMC Women's Health* 18(2018): 36. https://www.ncbi.nlm.nih .gov/pmc/articles/PMC5801702/.

Balsells, M., A. García-Patterson, and R. Corcov. "Systematic review and meta-analysis on the association of prepregnancy underweight and miscarriage." *European Journal of Obstetrics, Gynecology, and Reproductive Biology* 207 (December 2016): 73–79. https://www.ncbi.nlm.nih.gov/pubmed/27825031.

Banihani, S. A. "Effect of paracetamol on semen quality." *Andrologia* 50(1) (February 2018). https://www.ncbi.nlm.nih.gov/pubmed/28752572.

California Cryobank. "Sperm donor requirements." http://www.spermbank.com /how-it-works/sperm-donor-requirements.

Carlsen, E., A. M. Andersson, J. H. Petersen, and N. E. Skakkebaek. "History of febrile illness and variation in semen quality." *Human Reproduction* 18(10) (October 2003): 2089–92. https://www.ncbi.nlm.nih.gov/pubmed/14507826.

Carroll, K., A. M. Pottinger, S. Wynter, and V. DaCosta. "Marijuana use and its influence on sperm morphology and motility: Identified risk for fertility among Jamaican men." *Andrology* 8(1) (January 2020): 136–42. https://www.ncbi.nlm .nih.gov/pubmed/31267718.

Casilla-Lennon, M. M., S. Meltzer-Brody, and A. Z. Steiner. "The effect of anti-depressants on fertility." *American Journal of Obstetrics and Gynecology* 215(3) (September 2016): 314.e1–314.e5. doi:10.1016/j.ajog.2016.01.170. https://www.ncbi.nlm.nih.gov/pmc/articles/PMC4965341/.

Cavalcante, M. B., M. Sarno, A. B. Peixoto, E. Araujo Júnior, and R. Barini. "Obesity and recurrent miscarriage: A systematic review and meta-analysis." *Journal of Obstetrics and Gynaecology Research* 45(1) (January 2019): 30–38. https://www.ncbi .nlm.nih.gov/pubmed/30156037.

Centers for Disease Control and Prevention. "Antidepressant use among persons aged 12 and over: United States, 2011–2014." August 2017. https://www.cdc .gov/nchs/products/databriefs/db283.htm.

———. "Prevalence of obesity among adults and youth: United States, 2015–2016." NCHS Data Brief No. 288, October 2017. https://www.cdc.gov/nchs /products/databriefs/db288.htm.

———. "Smoking is down, but almost 38 million American adults still smoke." January 18, 2018. https://www.cdc.gov/media/releases/2018/p0118 -smoking-rates-declining.html.

———. "Trends in meeting the 2008 physical activity guidelines, 2008–2018 percentage." https://www.cdc.gov/physicalactivity/downloads/trends-in-the -prevalence-of-physical-activity-508.pdf.

Chiu, Y. H., M. C. Afeiche, A. J. Gaskins, P. L. Williams, J. Mendiola, N. Jørgensen, S. H. Swan, and J. E. Chavarro. "Sugar-sweetened beverage intake in relation to semen quality and reproductive hormone levels in young men." *Human Reproduction* 29(7) (July 2014): 1575–84. https://www.ncbi.nlm.nih.gov/pmc /articles/PMC4168308/.

Christou, M. A., P. A. Christou, G. Markozannes, A. Tsatsoulis, G. Mastorakos, and S. Tigas. "Effects of anabolic androgenic steroids on the reproductive system of athletes and recreational users: A systematic review and meta-analysis." *Sports Medicine* 47(9) (September 2017): 1869–83. https://www.ncbi.nlm .nih.gov/pubmed/28258581.

Cullen, K. A., A. S. Gentzke, M. D. Sawdey, J. T. Chang, G. M. Anic, T. W. Wang, M. R. Creamer, A. Jamal, B. K. Ambrose, and B. A. King. "E-cigarette use among youth in the United States, 2019." *JAMA* 322(21) (November 5, 2019): 2095–103. https://jamanetwork.com/journals/jama/fullarticle/2755265.

Dachille, G., M. Lamuraglia, M. Leone, A. Pagliarulo, G. Palasciano, M. T. Salerno, and G. M. Ludovico. "Erectile dysfunction and alcohol intake." *Urologia* 75(3) (July–September 2008): 170–76. https://www.ncbi.nlm.nih.gov /pubmed/21086346.

De Souza, M. J., A. Nattiv, E. Joy, M. Misra, N. I. Williams, R. J. Mallinson, J. C. Gibbs, M. Olmsted, M. Goolsby, and G. Matheson. "2014 Female Athlete Triad Coalition consensus statement on treatment and return to play of the female athlete triad: 1st International Conference held in San Francisco, California, May 2012, and 2nd International Conference held in Indianapolis, Indiana, May 2013." *British Journal of Sports Medicine* 48(4). https://bjsm.bmj.com /content/48/4/289.

Ding, J., X. Shang, Z. Zhang, H. Jing, J. Shao, Q. Fei, E. R. Rayburn, and H. Li. "FDA-approved medications that impair human spermatogenesis." *Oncotarget* 8(6) (February 7, 2017): 10714–25. https://www.ncbi.nlm.nih.gov/pmc /articles/PMC5354694/.

Dorey, G. "Is smoking a cause of erectile dysfunction? A literature review." *British Journal of Nursing* 10(7) (April 2001): 455–65. https://www.ncbi.nlm.nih.gov /pubmed/12070390.

Drobnis, E. Z., and A. K. Nangia. "Pain medications and male reproduction." *Advances in Experimental Medicine and Biology* 1034 (2017): 39–57. https://www.ncbi.nlm.nih.gov/pubmed/29256126.

Furukawa, S., T. Sakai, T. Niiya, H. Miyaoka, T. Miyake, S. Yamamoto, K. Maruyama et al. "Alcohol consumption and prevalence of erectile dysfunction in Japanese patients with type 2 diabetes mellitus: Baseline data from the Dogo Study." *Alcohol* 55 (September 2016): 17–22. https://www.ncbi.nlm.nih.gov/pubmed/27788774.

Gaskins, A. J., and J. E. Chavarro. "Diet and fertility: A review." *American Journal of Obstetrics and Gynecology* 218(4) (April 2018): 379–89. https://www.ncbi .nlm.nih.gov/pmc/articles/PMC5826784/.

Gaskins, A. J., M. C. Afeiche, R. Hauser, P. L. Williams, M. W. Gillman, C. Tanrikut, J. C. Petrozza, and J. E. Chavarro. "Paternal physical and sedentary activities in relation to semen quality and reproductive outcomes among couples from a fertility center." *Human Reproduction* 29(11) (November 2014): 2575–82. https://www.ncbi.nlm.nih.gov/pmc/articles/PMC4191451/.

Gaskins, A. J., J. W. Rich-Edwards, P. L. Williams, T. L. Toth, S. A. Missmer, and J. E. Chavarro. "Pre-pregnancy caffeine and caffeinated beverage intake and risk of spontaneous abortion." *European Journal of Nutrition* 57(1) (February 2018): 107–17. https://www.ncbi.nlm.nih.gov/pmc/articles/PMC5332346/.

———. "Prepregnancy Low to Moderate Alcohol Intake Is Not Associated with Risk of Spontaneous Abortion or Stillbirth." *Journal of Nutrition* 146(4) (April 2016): 799–805. https://www.ncbi.nlm.nih.gov/pmc/articles/PMC4807650/.

Gebreegziabher, Y., E. Marcos, W. McKinon, and G. Rogers. "Sperm characteristics of endurance trained cyclists." *International Journal of Sports Medicine* 25(4) (May 2004): 247–51. https://www.ncbi.nlm.nih.gov/pubmed/15162242.

Gollenberg, A. L., F. Liu, C. Brazil, E. Z. Drobnis, D. Guzick, J. W. Overstreet, J. B. Redmon, A. Sparks, C. Wang, and S. H. Swan. "Semen quality in fertile men in relation to psychosocial stress." *Fertility and Sterility* 93(4) (March 1, 2010): 1104–11. https://www.ncbi.nlm.nih.gov/pubmed/19243749.

Grant, B. F., S. P. Chou, T. D. Saha, R. P. Pickering, B. T. Kerridge, W. J. Ruan, B. Huang et al. "Prevalence of 12-month alcohol use, high-risk drinking, and *DSM-IV* alcohol use disorder in the United States, 2001–2002 to 2012–2013." *JAMA Psychiatry* 74(9) (2017): 911–23. https://jamanetwork.com/journals/jamapsychiatry/fullarticle/2647079.

Gundersen, T. D., N. Jørgensen, A. M. Andersson, A. K. Bang, L. Nordkap, N. E. Skakkebæk, L. Priskorn, A. Juul, and T. K. Jensen. "Association between use of marijuana and male reproductive hormones and semen quality: A study among 1,215 healthy young men." *American Journal of Epidemiology* 182(6) (August 16, 2015): 473–81. https://www.ncbi.nlm.nih.gov/pubmed/26283092.

Hamzelou, J. "Weird cells in your semen? Don't panic, you might just have flu." *New Scientist*, June 30, 2015. https://www.newscientist.com/article/dn27809-weird-cells-in-your-semen-dont-panic-you-might-just-have-flu/.

Hawkins Bressler, L., L. A. Bernardi, P. J. De Chavez, D. D. Baird, M. R. Carnethon, and E. E. Marsh. "Alcohol, cigarette smoking, and ovarian reserve in reproductive-age African-American women." *American Journal of Obstetrics and Gynecology* 215(6) (December 2016): 758.e1–758.e9. https://www.ncbi.nlm.nih.gov/pmc/articles/PMC5124512/.

Hyland, A., K. M. Piazza, K. M. Hovey, J. K. Ockene, C. A. Andrews, C. Rivard, and J. Wactawski-Wende. "Associations of lifetime active and passive smoking with spontaneous abortion, stillbirth and tubal ectopic pregnancy: A cross-sectional analysis of historical data from the Women's Health Initiative." *Tobacco Control* 24(4) (July 2015): 328–35. https://www.ncbi.nlm.nih.gov/pubmed/24572626.

Hyland, A., K. Piazza, K. M. Hovey, H. A. Tindle, J. E. Manson, C. Messina, C. Rivard, D. Smith, and J. Wactawski-Wende. "Associations between lifetime tobacco exposure with infertility and age at natural menopause: The Women's Health Initiative Observational Study." *Tobacco Control* 25(6) (November 2016): 706–14. https://www.ncbi.nlm.nih.gov/pubmed/26666428.

Ippolito, A. C., A. D. Seelig, T. M. Powell, A. M. S. Conlin, N. F. Crum-Cianflone, H. Lemus, C. J. Sevick, and C. A. LeardMann. "Risk factors associated with miscarriage and impaired fecundity among United States servicewomen during the recent conflicts in Iraq and Afghanistan." *Women's Health Issues* 27(3) (May–June 2017): 356–65. https://www.ncbi.nlm.nih.gov/pubmed/28160994.

Jensen, T. K., M. Gottschau, J. O. B. Madsen, A-M. Andersson, T. H. Lassen, N. E. Skakkebaek, S. H. Swan, L. Priskorn, A. Juul, and N. Jørgensen. "Habitual alcohol consumption associated with reduced semen quality and changes in reproductive hormones: A cross-sectional study among 1,221 young Danish men." *BMJ Open* 4(9) (2014): e005462. https://www.ncbi.nlm.nih.gov/pmc/articles/PMC4185337/.

Lania, A., L. Gianotti, I. Gagliardi, M. Bondanelli, W. Vena, and M. R. Ambrosio. "Functional hypothalamic and drug-induced amenorrhea: An overview." *Journal of Endocrinological Investigation* 42(9) (September 2019): 1001–10. https://www.ncbi.nlm.nih.gov/pubmed/30742257.

Luque, E. M., A. Tissera, M. P. Gaggino, R. I. Molina, A. Mangeaud, L. M. Vincenti, F. Beltramone et al. "Body mass index and human sperm quality: Neither one extreme nor the other." *Reproduction, Fertility, and Development* 29(4) (April 2017): 731–39. https://www.ncbi.nlm.nih.gov/pubmed/26678380.

Millett, C., L. M. Wen, C. Rissel, A. Smith, J. Richters, A. Grulich, and R. de Visser. "Smoking and erectile dysfunction: Findings from a representative sample of Australian men." *Tobacco Control* 15(2) (April 2006): 136–39. https://www.ncbi .nlm.nih.gov/pmc/articles/PMC2563576/.

Mulligan, T., M. F. Frick, Q. C. Zuraw, A. Stemhagen, and C. McWhirter. "Prevalence of hypogonadism in males aged at least 45 years: The HIM study." *International Journal of Clinical Practice* 60(7) (July 2006): 762–69. https://www.ncbi.nlm.nih .gov/pmc/articles/PMC1569444/.

Nagma, S., G. Kapoor, R. Bharti, A. Batra, A. Batra, A. Aggarwal, and A. Sablok. "To evaluate the effect of perceived stress on menstrual function." *Journal of Clinical and Diagnostic Research* 9(3) (March 2015): QC01–QC03. https://www.ncbi .nlm.nih.gov/pmc/articles/PMC4413117/.

Nassan, F. L., M. Arvizu, L. Mínguez-Alarcón, A. J. Gaskins, P. L. Williams, J. C. Petrozza, R. Hauser, J. E. Chavarro, and EARTH Study Team. "Marijuana smoking and outcomes of infertility treatment with assisted reproductive technologies." *Human Reproduction* 34(9) (September 29, 2019): 1818–29. https://www.ncbi.nlm .nih.gov/pubmed/31505640.

National Institute on Drug Abuse. "What is the scope of marijuana use in the United States?" Last updated December 2019. https://www.drugabuse.gov/publica tions/research-reports/marijuana/what-scope-marijuana-use-in-united-states.

Nordkap, L., T. K. Jensen, A. M. Hansen, T. H. Lassen, A. K. Bang, U. N. Joensen, M. Blomberg Jensen, N. E. Skakkebaek, and N. Jørgensen. "Psychological stress and testicular function: A cross-sectional study of 1,215 Danish men." *Fertility and Sterility* 105(1) (January 2016): 174–87. https://www.ncbi.nlm.nih.gov /pubmed/26477499.

NW Cryobank. "NW Cryobank sperm donor requirements." https://www.nwsperm .com/how-it-works/sperm-donor-requirements.

Office on Women's Health. "Weight, fertility, and pregnancy." Page last updated December 27, 2018. https://www.womenshealth.gov/healthy-weight/weight -fertility-and-pregnancy.

Palermo, G. D., Q. V. Neri, T. Cozzubbo, S. Cheung, N. Pereira, and Z. Rosenwaks. "Shedding light on the nature of seminal round cells." *PLoS One* 11(3) (March 16, 2016): e0151640. https://journals.plos.org/plosone/article?id=10.1371 /journal.pone.0151640.

Panara, K., J. M. Masterson, L. F. Savio, and R. Ramasamy. "Adverse effects of common sports and recreational activities on male reproduction." *European Urology Focus* 5(6) (November 2019): 1146–51. https://www.ncbi.nlm.nih.gov/pubmed/29731401.

Patra, P. B., and R. M. Wadsworth. "Quantitative evaluation of spermatogenesis in mice following chronic exposure to cannabinoids." *Andrologia* 23(2) (March–April 1991): 151–56. https://www.ncbi.nlm.nih.gov/pubmed/1659250.

Priskorn, L., T. K. Jensen, A. K. Bang, L. Nordkap, U. N. Joensen, T. H. Lassen, I. A. Olesen, S. H. Swan, N. E. Skakkebaek, and N. Jørgensen. "Is sedentary lifestyle associated with testicular function? A cross-sectional study of 1,210 men." *American Journal of Epidemiology* 184(4) (August 15, 2016): 284–94. https://www.ncbi.nlm.nih.gov/pubmed/27501721.

Qu, F., Y. Wu, Y-H. Zhu, J. Barry, T. Ding, G. Baio, R. Muscat, B. K. Todd, F-F. Wang, and P. J. Hardiman. "The association between psychological stress and miscarriage: A systematic review and meta-analysis." *Scientific Reports* 7 (May 2017): 1731. https://www.ncbi.nlm.nih.gov/pmc/articles/PMC5431920/.

Radwan, M., J. Jurewicz, D. Merecz-Kot, W. Sobala, P. Radwan, M. Bochenek, and W. Hanke. "Sperm DNA damage—the effect of stress and everyday life factors." *International Journal of Impotence Research* 28(4) (July 2016): 148–54. https://www.ncbi.nlm.nih.gov/pubmed/27076112.

Rahali, D., A. Jrad-Lamine, Y. Dallagi, Y. Bdiri, N. Ba, M. El May, S. El Fazaa, and N. El Golli. "Semen parameter alteration, histological changes and role of oxidative stress in adult rat epididymis on exposure to electronic cigarette refill liquid." *Chinese Journal of Physiology* 61(2) (April 30, 2018): 75–84. https://www.ncbi.nlm.nih.gov/pubmed/29526076.

Ramaraju, G. A., S. Teppala, K. Prathigudupu, M. Kalagara, S. Thota, M. Kota, and R. Cheemakurthi. "Association between obesity and sperm quality." *Andrologia* 50(3) (April 2018). https://www.ncbi.nlm.nih.gov/pubmed/28929508.

Remes, O., C. Brayne, R. van der Linde, and L. Lafortune. "A systematic review of reviews on the prevalence of anxiety disorders in adult populations." *Brain and Behavior* 6(7) (July 2016): e00497. https://onlinelibrary.wiley.com/doi/full/10.1002/brb3.497.

Ricci, E., S. Noli, S. Ferrari, I. La Vecchia, S. Cipriani, V. De Cosmi, E. Somigliana, and F. Parazzini. "Alcohol intake and semen variables: Cross-sectional analysis of a prospective cohort study of men referring to an Italian fertility clinic." *Andrology* 6(5) (September 2018): 690–96. https://www.ncbi.nlm.nih.gov/pubmed/30019500.

Santillano, V. "Is height advantage a tall tale?" *More.* Updated December 27, 2009. https://www.more.com/lifestyle/exercise-health/height-advantage-tall-tale/.

Schlossberg, M. "5 Things you need to know about whiskey dick, the greatest curse known to mankind." *Men's Health*, September 21, 2017. https://www.menshealth

.com/sex-women/a19535862/whiskey-dick-is-real-and-heres-the-science
-behind-it/.

Schuel, H., R. Schuel, A. M. Zimmerman, and S. Zimmerman. "Cannabinoids
reduce fertility of sea urchin sperm." *Biochemistry and Cell Biology* 65(2)
(February 1987): 130–36. https://www.ncbi.nlm.nih.gov/pubmed/3030370.

Sharma, R., A. Harley, A. Agarwal, and S. C. Esteves. "Cigarette smoking and semen
quality: A new meta-analysis examining the effect of the 2010 World Health
Organization laboratory methods for the examination of human semen." *European Urology* 70(4) (October 2016): 635–45. https://www.ncbi.nlm.nih.gov
/pubmed/27113031.

Sperm Bank of California. "How to qualify as a sperm donor?" https://www.the
spermbankofca.org/content/how-qualify-sperm-donor.

Swan, S. H., F. Liu, J. W. Overstreet, C. Brazil, and N. E. Skakkebaek. "Semen quality
of fertile US males in relation to their mothers' beef consumption during pregnancy." *Human Reproduction* 22(6) (June 2007): 1497–1502. https://www.ncbi
.nlm.nih.gov/pubmed/17392290.

Tatem, A. J., J. Beilan, J. R. Kovac, and L. I. Lipshultz. "Management of anabolic
steroid-induced infertility: Novel strategies for fertility maintenance and recovery." *World Journal of Men's Health* 38(2) (April 2020): 141–50. https://wjmh.org
/DOIx.php?id=10.5534/wjmh.190002.

Chapter Seven:
Silent, Ubiquitous Threats: *The Dangers of Plastics and Modern Chemicals*

Barrett, E. S., S. Sathyanarayana, O. Mbowe, S. W. Thurston, J. B. Redmon, R. H. N.
Nguyen, and S. H. Swan. "First-trimester urinary bisphenol A concentration in
relation to anogenital distance, an androgen-sensitive measure of reproductive
development, in infant girls." *Environmental Health Perspectives*, July 11, 2017.
https://ehp.niehs.nih.gov/doi/10.1289/EHP875.

Barrett, E. S., and M. Sobolewski. "Polycystic ovary syndrome: Do endocrine disrupting chemicals play a role?" *Seminars in Reproductive Medicine* 32(3) (May
2014): 166–76. https://www.ncbi.nlm.nih.gov/pmc/articles/PMC4086778/.

Bienkowski, B. " 'Environmentally friendly' flame retardants break down into potentially toxic chemicals." *Environmental Health News*, January 9, 2019. https://
www.ehn.org/environmentally-friendly-flame-retardants-break-down-into
-potentially-toxic-chemicals-2625440344.html.

Bloom, M. S., B. W. Whitcomb, Z. Chen, A. Ye, K. Kannan, and G. M. Buck Louis.
"Associations between urinary phthalate concentrations and semen quality parameters in a general population." *Human Reproduction* 30(11) (September 2015): 2645–
57. https://www.ncbi.nlm.nih.gov/pmc/articles/PMC4605371/pdf/dev219.pdf.

Bornehag, C. G., F. Carlstedt, B. A. Jönsson, C. H. Lindh, T. K. Jensen, A. Bodin, C. Jonsson, S. Janson, and S. H. Swan. "Prenatal phthalate exposures and anogenital distance in Swedish boys." *Environmental Health Perspectives* 123(1) (January 2015): 101–7. https://www.ncbi.nlm.nih.gov/pmc/articles/PMC4286276/.

Bretveld, R., G. A. Zielhuis, and N. Roeleveld. "Time to pregnancy among female greenhouse workers." *Scandinavian Journal of Work, Environment, & Health* 32(5) (October 2006): 359–67. https://www.ncbi.nlm.nih.gov/pubmed/17091203.

Carson, R. *Silent Spring.* Boston: Houghton Mifflin, 1962.

Caserta, D., N. Di Segni, M. Mallozzi, V. Giovanale, A. Mantovani, R. Marci, and M. Moscarini. "Bisphenol A and the female reproductive tract: An overview of recent laboratory evidence and epidemiological studies." *Reproductive Biology and Endocrinology* 12 (2014): 37. https://www.ncbi.nlm.nih.gov/pmc/articles /PMC4019948/.

Centers for Disease Control and Prevention. "National report on human exposure to environmental chemicals." Updated tables, January 2019. https://www.cdc.gov /exposurereport/index.html.

Chevrier, C., C. Warembourg, E. Gaudreau, C. Monfort, A. Le Blanc, L. Guldner, and S. Cordier. "Organochlorine pesticides, polychlorinated biphenyls, seafood consumption, and time-to-pregnancy." *Epidemiology* 24(2) (March 2013): 251–60. https://www.ncbi.nlm.nih.gov/pubmed/23348067.

Choi, G., Y. B. Wang, R. Sundaram, Z. Chen, D. B. Barr, G. M. Buck Louis, and M. M. Smarr. "Polybrominated diphenyl ethers and incident pregnancy loss: The LIFE Study." *Environmental Research* 168 (January 2019): 375–81. https://www.ncbi.nlm.nih.gov/pmc/articles/PMC6294303/.

Collaborative on Health and the Environment. "Regrettable replacements: The next generation of endocrine disrupting chemicals." October 24, 2017. https://www.healthand environment.org/partnership_calls/95948.

Condorelli, R., A. E. Calogero, and S. La Vignera. "Relationship between testicular volume and conventional or nonconventional sperm parameters." *International Journal of Endocrinology*, 2013. Article ID 145792. https://www.hindawi.com /journals/ije/2013/145792/.

Di Nisio, A., I. Sabovic, U. Valente, S. Tescari, M. S. Rocca, D. Guidolin, S. Dall'Acqua et al. "Endocrine disruption of androgenic activity by perfluoroalkyl substances: Clinical and experimental evidence." *Journal of Clinical Endocrinology and Metabolism* 104(4) (April 1, 2019): 1259–71. https://www.ncbi.nlm.nih.gov/pubmed/30403786.

"The dose makes the poison." ChemistrySafetyFacts.org. https://www.chemicalsafe tyfacts.org/dose-makes-poison-gallery/.

Eskenazi, B., P. Mocarelli, M. Warner, S. Samuels, P. Vercellini, D. Olive, L. L. Needham et al. "Serum dioxin concentrations and endometriosis: A cohort study in Seveso, Italy." *Environmental Health Perspectives* 110(7) (July 2002): 629–34. https://www.ncbi.nlm.nih.gov/pmc/articles/PMC1240907/.

Eskenazi, B., M. Warner, A. R. Marks, S. Samuels, L. Needham, P. Brambilla, and P. Mocarelli. "Serum dioxin concentrations and time to pregnancy." *Epidemiology* 21(2) (March 2010): 224–31. https://www.ncbi.nlm.nih.gov/pmc/articles/PMC6267871/.

Harley, K. G., A. R. Marks, J. Chevrier, A. Bradman, A. Sjödin, and B. Eskenazi. "PBDE concentrations in women's serum and fecundability." *Environmental Health Perspectives* 118(5) (May 2010): 699–704. https://www.ncbi.nlm.nih.gov/pmc/articles/PMC2866688/.

Harley, K. G., S. A. Rauch, J. Chevrier, K. Kogut, K. L. Parra, C. Trujillo, R. H. Lustig et al. "Association of prenatal and childhood PBDE exposure with timing of puberty in boys and girls." *Environment International* 100 (March 2017): 132–38. https://www.ncbi.nlm.nih.gov/pmc/articles/PMC5308219/.

Hart, R. J., H. Frederiksen, D. A. Doherty, J. A. Keelan, N. E. Skakkebaek, N. S. Minaee, R. McLachlan et al. "The possible impact of antenatal exposure to ubiquitous phthalates upon male reproductive function at 20 years of age." *Frontiers in Endocrinology* 9 (June 2018): 288. https://www.ncbi.nlm.nih.gov/pmc/articles/PMC5996240/.

Herrero, Ó., M. Aquilino, P. Sánchez-Argüello, and R. Planelló. "The BPA-substitute bisphenol S alters the transcription of genes related to endocrine, stress response and biotransformation pathways in the aquatic midge *Chironomus riparius* (Diptera, Chironomidae)." *PLoS One* 13(2) (2018): e0193387. https://www.ncbi.nlm.nih.gov/pmc/articles/PMC5821402/.

Hormone Health Network. "Endocrine-disrupting chemicals (EDCs)." https://www.hormone.org/your-health-and-hormones/endocrine-disrupting-chemicals-edcs.

Houlihan, J., C. Brody, and B. Schwan. "Not too pretty: Phthalates, beauty products & the FDA." Environmental Working Group, July 2002. https://www.safecosmetics.org/wp-content/uploads/2015/02/Not-Too-Pretty.pdf.

Hu, Y., L. Ji, Y. Zhang, R. Shi, W. Han, L. A. Tse, R. Pan et al. "Organophosphate and pyrethroid pesticide exposures measured before conception and associations with time to pregnancy in Chinese couples enrolled in the Shanghai Birth Cohort." *Environmental Health Perspectives* 126(7) (July 9, 2018): 077001. https://www.ncbi.nlm.nih.gov/pmc/articles/PMC6108871/.

Kandaraki, E., A. Chatzigeorgiou, S. Livadas, E. Palioura, F. Economou, M. Koutsilieris, S. Palimeri, D. Panidis, and E. Diamanti-Kandarakis. "Endocrine disruptors and polycystic ovary syndrome (PCOS): Elevated serum levels of bisphenol A in women with PCOS." *Journal of Clinical Endocrinology and Metabolism* 96(3) (March 2011): E480–E484. https://academic.oup.com/jcem/article/96/3/E480/2597282.

Lathi, R. B., C. A. Liebert, K. F. Brookfield, J. A. Taylor, F. S. vom Saal, V. Y. Fujimoto, and V. L. Baker. "Conjugated bisphenol A (BPA) in maternal serum in relation to miscarriage risk." *Fertility and Sterility* 102(1) (July 2014): 123–28. https://www.ncbi.nlm.nih.gov/pmc/articles/PMC4711263/.

Li, D. K., Z. Zhou, M. Miao, Y. He, J. T. Wang, J. Ferber, L. J. Herrinton, E. S. Gao, and W. Yuan. "Urine bisphenol-A (BPA) level in relation to semen quality." *Fertility and Sterility* 95(2) (February 2011): 625–30. https://www.science direct.com/science/article/abs/pii/S0015028210025872.

MacKendrick, N. *Better Safe Than Sorry.* Oakland: University of California Press, 2018.

Miao, M., W. Yuan, Y. He, Z. Zhou, J. Wang, E. Gao, G. Li, and D. K. Li. "In utero exposure to bisphenol-A and anogenital distance of male offspring." *Birth Defects Research Part A: Clinical and Molecular Teratology* 91(10) (October 2011): 867–72. https://pubmed.ncbi.nlm.nih.gov/21987463/.

"Microplastics found in human stools for first time." Technology Networks, October 23, 2018. https://www.technologynetworks.com/applied-sciences/news /microplastics-found-in-human-stools-for-first-time-310862.

Mínguez-Alarcón, L., O. Sergeyev, J. S. Burns, P. L. Williams, M. M. Lee, S. A. Korrick, L. Smigulina, B. Revich, and R. Hauser. "A longitudinal study of peripubertal serum organochlorine concentrations and semen parameters in young men: The Russian Children's Study." *Environmental Health Perspectives* 125(3) (March 2017): 160–466. https://www.ncbi.nlm.nih.gov/pmc/articles/PMC5332179/.

Mitro, S. D., R. E. Dodson, V. Singla, G. Adamkiewicz, A. F. Elmi, M. K. Tilly, and A. R. Zota. "Consumer product chemicals in indoor dust: A quantitative meta-analysis of U.S. studies." *Environmental Science & Technology* 50(19) (October 4, 2016): 10661–72. https://www.ncbi.nlm.nih.gov/pmc/articles /PMC5052660/.

National Pesticide Information Center. "Pesticides—what's my risk?" Last updated April 11, 2012. http://npic.orst.edu/factsheets/WhatsMyRisk.html.

Nevoral, J., Y. Kolinko, J. Moravec, T. Žalmanová, K. Hošková, Š. Prokešová, P. Klein et al. "Long-term exposure to very low doses of bisphenol S affects female reproduction." *Reproduction* 156(1) (July 2018): 47–57. https://www.ncbi.nlm .nih.gov/pubmed/29748175.

Özel, S., A. Tokmak, O. Aykut, A. Aktulay, N. Hançerlioğullari, and Y. Engin Ustun. "Serum levels of phthalates and bisphenol-A in patients with primary ovarian insufficiency." *Gynecological Endocrinology* 35(4) (April 2019): 364–67. https:// www.ncbi.nlm.nih.gov/pubmed/30638094.

Planned Parenthood. "Sexual and reproductive anatomy." https://www.planned parenthood.org/learn/health-and-wellness/sexual-and-reproductive-anatomy.

Radke, E. G., J. M. Braun, J. D. Meeker, and G. S. Cooper. "Phthalate exposure and male reproductive outcomes: A systematic review of the human epidemiological evidence." *Environment International* 121 (pt. 1) (December 2018): 764–93. https://www.sciencedirect.com/science/article/pii/S0160412018303404.

Rafizadeh, D. "BPA-free isn't always better: The dangers of BPS, a BPA substitute." *Yale Scientific*, August 17, 2016. http://www.yalescientific.org/2016/08 /bpa-free-isnt-always-better-the-dangers-of-bps-a-bpa-substitute/.

Ratcliffe, J. M., S. M. Schrader, K. Steenland, D. E. Clapp, T. Turner, and R. W. Hornung. "Semen quality in papaya workers with long term exposure to ethylene dibromide." *British Journal of Industrial Medicine* 44(5) (May 1987): 317–26. https://www.ncbi.nlm.nih.gov/pmc/articles/PMC1007829/.

Rutkowska, A. Z., and E. Diamanti-Kandarakis. "Polycystic ovary syndrome and environmental toxins." *Fertility and Sterility* 106(4) (September 15, 2016): 948–58. https://www.ncbi.nlm.nih.gov/pubmed/27559705.

Smith, R., and B. Lourie. *Slow Death by Rubber Duck: How the Toxicity of Everyday Life Affects Our Health.* Toronto: Knopf Canada, expanded, updated edition, 2019.

Stoiber, T. "Study: Banned since 2004, toxic flame retardants persist in U.S. newborns." Environmental Working Group, July 11, 2017. https://www.ewg.org /enviroblog/2017/07/study-banned-2004-toxic-flame-retardants-persist-us -newborns.

Swan, S. H., R. L. Kruse, F. Liu, D. B. Barr, E. Z. Drobnis, J. B. Redmon, C. Wang, C. Brazil, and J. W. Overstreet. "Semen quality in relation to biomarkers of pesticide exposure." *Environmental Health Perspectives* 111(12) (September 2003): 1478–84. https://www.ncbi.nlm.nih.gov/pmc/articles/PMC1241650/.

Toft, G., A. M. Thulstrup, B. A. Jönsson, H. S. Pedersen, J. K. Ludwicki, V. Zvezday, and J. P. Bonde. "Fetal loss and maternal serum levels of 2,2',4,4',5,5'hexachlorbiphenyl (CB-153) and 1,1-dichloro-2,2-bis(p-chlorophenyl)ethylene (p,p'-DDE) exposure: A cohort study in Greenland and two European populations." *Environmental Health* 9 (2010): 22. https://www.ncbi.nlm.nih.gov/pmc /articles/PMC2877014/.

Toumi, K., L. Joly, C. Vleminckx, and B. Schiffers. "Risk assessment of florists exposed to pesticide residues through handling of flowers and preparing bouquets." *International Journal of Environmental Research and Public Health* 14(5) (May 2017): 526. https://www.ncbi.nlm.nih.gov/pmc/articles/PMC5451977/.

Vabre, P., N. Gatimel, J. Moreau, V. Gayrard, N. Picard-Hagen, J. Parinaud, and R. D. Leandri. "Environmental pollutants, a possible etiology for premature ovarian insufficiency: A narrative review of animal and human data." *Environmental Health* 16(37) (2017). https://www.ncbi.nlm.nih.gov/pmc/articles /PMC5384040/.

Vandenberg, L. N., T. Colborn, T. B. Hayes, J. J. Heindel, D. R. Jacobs Jr., D-H. Lee, T. Shioda et al. "Hormones and endocrine-disrupting chemicals: Low-dose effects and nonmonotonic dose responses." *Endocrine Reviews* 33(3) (June 2012): 378–455. https://www.ncbi.nlm.nih.gov/pmc/articles/PMC3365860/.

Vogel, S. A. "The politics of plastics: The making and unmaking of bisphenol

A 'safety.'" *American Journal of Public Health* 99(S3) (2009): S559–S566. https://www.ncbi.nlm.nih.gov/pmc/articles/PMC2774166/.

Zhang, J., L. Chen, L. Xiao, F. Ouyang, Q. Y. Zhang, and Z. C. Luo. "Polybrominated diphenyl ether concentrations in human breast milk specimens worldwide." *Epidemiology* 28(suppl. 1) (October 2017): S89–S97. https://www.ncbi .nlm.nih.gov/pubmed/29028681.

Ziv-Gal, A., and J. A. Flaws. "Evidence for bisphenol A–induced female infertility: Review (2007–2016)." *Fertility and Sterility* 106(4) (September 15, 2016): 827–56. https://www.ncbi.nlm.nih.gov/pmc/articles/PMC5026908/.

Ziv-Gal, A., L. Gallicchio, C. Chiang, S. N. Ther, S. R. Miller, H. A. Zacur, R. L. Dills, and J. A. Flaws. "Phthalate metabolite levels and menopausal hot flashes in midlife women." *Reproductive Toxicology* 60 (April 2016): 76–81. https://www.ncbi.nlm.nih.gov/pmc/articles/PMC4867120/.

Chapter Eight:
The Long Reach of Exposures: *Reproductive Ripple Effects*

Brown, A. S., and E. S. Susser. "Prenatal nutritional deficiency and risk of adult schizophrenia." *Schizophrenia Bulletin* 34(6) (November 2008): 1054–63. https://www.ncbi.nlm.nih.gov/pmc/articles/PMC2632499/.

Bygren, L. O., P. Tinghög, J. Carstensen, S. Edvinsson, G. Kaati, M. E. Pembrey, and M. Sjöström. "Change in paternal grandmothers' early food supply influenced cardiovascular mortality of the female grandchildren." *BMC Genetics* 15 (February 2014): 12. https://www.ncbi.nlm.nih.gov/pmc/articles /PMC3929550/.

Cedars, M. I., S. E. Taymans, L. V. DePaolo, L. Warner, S. B. Moss, and M. L. Eisenberg. "The sixth vital sign: What reproduction tells us about overall health. Proceedings from a NICHD/CDC Workshop." *Human Reproduction Open* 2017(2) (2017). https://www.ncbi.nlm.nih.gov/pmc/articles/PMC6276647/.

Charalampopoulos, D., A. McLoughlin, C. E. Elks, and K. K. Ong. "Age at menarche and risks of all-cause and cardiovascular death: A systematic review and meta-analysis." *American Journal of Epidemiology* 180(1) (July 2014): 29–40. https://www.ncbi.nlm.nih.gov/pmc/articles/PMC4070937/.

Dolinoy, D. C., D. Huang, and R. L. Jirtle. "Maternal nutrient supplementation counteracts bisphenol A–induced DNA hypomethylation in early development." *Proceedings of the National Academy of Sciences* 104(32) (August 2007): 13056–61. https://www.ncbi.nlm.nih.gov/pmc/articles/PMC1941790/.

Eisenberg, M. L., S. Li, B. Behr, M. R. Cullen, D. Galusha, D. J. Lamb, and L. I. Lipshultz. "Semen quality, infertility and mortality in the USA." *Human Reproduction* 29(7) (July 2014): 1567–74. https://www.ncbi.nlm.nih.gov/pmc/articles /PMC4059337/pdf/deu106.pdf.

Eisenberg, M. L., S. Li, M. R. Cullen, and L. C. Baker. "Increased risk of incident chronic medical conditions in infertile men: Analysis of United States claims data." *Fertility and Sterility* 105(3) (March 2016): 629–36. https://www.ncbi .nlm.nih.gov/pubmed/26674559.

Elias, S. G., P. A. H. van Noord, P. H. M. Peeters, I. D. Tonkelaar, and D. E. Grobbee. "Caloric restriction reduces age at menopause: The effect of the 1944–1945 Dutch famine." *Menopause* 25(11) (November 2018): 1232–37. https://www.ncbi.nlm.nih .gov/pubmed/30358718.

Hatipoğlu, N., and S. Kurtoğlu. "Micropenis: Etiology, diagnosis and treatment approaches." *Journal of Clinical Research in Pediatric Endocrinology* 5(4) (December 2013): 217–23. https://www.ncbi.nlm.nih.gov/pmc/articles /PMC3890219/.

Jensen, T. K., R. Jacobsen, K. Christensen, N. C. Nielsen, and E. Bostofte. "Good semen quality and life expectancy: A cohort study of 43,277 men." *American Journal of Epidemiology* 170(5) (September 2009): 559–65. https://www.ncbi .nlm.nih.gov/pubmed/19635736.

Kanherkar, R. R., N. Bhatia-Dey, and A. B. Csoka. "Epigenetics across the human lifespan." *Frontiers in Cell Developmental Biology* 2(49) (September 9, 2014). https://www.frontiersin.org/articles/10.3389/fcell.2014.00049/full.

Ly, L., D. Chan, M. Aarabi, M. Landry, N. A. Behan, A. J. MacFarlane, and J. Trasler. "Intergenerational impact of paternal lifetime exposures to both folic acid deficiency and supplementation on reproductive outcomes and imprinted gene methylation." *Molecular Human Reproduction* 23(7) (July 2017): 461–77. https://www.ncbi.nlm.nih.gov/pmc/articles/PMC5909862/.

MacMahon, B., P. Cole, T. M. Lin, C. R. Lowe, A. P. Mirra, B. Ravnihar, E. J. Salber, V. G. Valaoras, and S. Yuasa. "Age at first birth and breast cancer risk." *Bulletin of the World Health Organization* 43(2) (1970): 209–21. https://www.ncbi.nlm .nih.gov/pmc/articles/PMC2427645/.

Menezo, Y., B. Dale, and K. Elder. "The negative impact of the environment on methylation/epigenetic marking in gametes and embryos: A plea for action to protect the fertility of future generations." *Molecular Reproduction & Development* 86(10) (October 2019): 1273–82. https://www.ncbi.nlm.nih.gov /pubmed/30653787.

"Menstruation and breastfeeding." La Leche League International. https://www.llli .org/breastfeeding-info/menstruation/.

Mørkve Knudsen, T., F. I. Rezwan, Y. Jiang, W. Karmaus, C. Svanes, and J. W. Holloway. "Transgenerational and intergenerational epigenetic inheritance in allergic diseases." *Journal of Allergy and Clinical Immunology* 142(3) (September 2018): 765–72. https://www.ncbi.nlm.nih.gov/pmc/articles/PMC6167012/.

Murugappan, G., S. Li, R. B. Lathi, V. L. Baker, and M. L. Eisenberg. "Risk of cancer in infertile women: Analysis of US claims data." *Human Reproduction* 34(5)

(May 1, 2019): 894–902. https://www.ncbi.nlm.nih.gov/pubmed/30863841.

Myers, P. "Science: Are we in a male fertility death spiral?" *Environmental Health News*, July 26, 2017. https://www.ehn.org/science_are_we_in_a_male_fertility_death_spiral-2497202098.html.

Nilsson, E. E., I. Sadler-Riggleman, and M. K. Skinner. "Environmentally induced epigenetic transgenerational inheritance of disease." *Environmental Epigenetics* 4(2) (April 2018): 1–13. https://www.ncbi.nlm.nih.gov/pmc/articles/PMC6051467/.

Northstone, K., J. Golding, G. Davey Smith, L. L. Miller, and M. Pembrey. "Prepubertal start of father's smoking and increased body fat in his sons: Further characterisation of paternal transgenerational responses." *European Journal of Human Genetics* 22(12) (December 2014): 1382–86. https://www.ncbi.nlm.nih.gov/pmc/articles/PMC4085023/.

Painter, R. C., C. Osmond, P. Gluckman, M. Hanson, D. I. W. Phillips, and T. J. Roseboom. "Transgenerational effects of prenatal exposure to the Dutch famine on neonatal adiposity and health in later life." *BJOG* 115(10) (September 2008): 1243–49. https://obgyn.onlinelibrary.wiley.com/doi/full/10.1111/j.1471-0528.2008.01822.x.

Palmer, J. R., A. L. Herbst, K. L. Noller, D. A. Boggs, R. Troisi, L. Titus-Ernstoff, E. E. Hatch, L. A. Wise, W. C. Strohsnitter, and R. N. Hoover. "Urogenital abnormalities in men exposed to diethylstilbestrol *in utero*: A cohort study." *Environmental Health* 8(37) (August 2009). https://www.ncbi.nlm.nih.gov/pmc/articles/PMC2739506/.

Pembrey, M. E., L. O. Bygren, G. Kaati, S. Edvinsson, K. Northstone, M. Sjöström, J. Golding, and ALSPAC Study Team. "Sex-specific, male-line transgenerational responses in humans." *European Journal of Human Genetics* 14(2) (February 2006): 159–66. https://www.ncbi.nlm.nih.gov/pubmed/16391557.

Rodgers, A. B., and T. L. Bale. "Germ cell origins of PTSD risk: The transgenerational impact of parental stress experience." *Biological Psychiatry* 78(5) (September 1, 2015): 307–14. https://www.ncbi.nlm.nih.gov/pmc/articles/PMC4526334/.

Rodgers, A. B., C. P. Morgan, S. L. Bronson, S. Revello, and T. L. Bale. "Paternal stress exposure alters sperm microRNA content and reprograms offspring HPA stress axis regulation." *Journal of Neuroscience* 33(21) (May 22, 2013): 9003–12. https://www.ncbi.nlm.nih.gov/pmc/articles/PMC3712504/.

Schulz, L. C. "The Dutch Hunger Winter and the developmental origins of health and disease." *Proceedings of the National Academy of Sciences* 107(39) (September 28, 2010): 16757–58. https://www.ncbi.nlm.nih.gov/pmc/articles/PMC2947916/.

Tournaire, M. D., E. Devouche, S. Epelboin, A. Cabau, A. Dunbavand, and A. Levadou. "Birth defects in children of men exposed in utero to diethylstilbestrol (DES)." *Therapie* 73(5) (October 2018): 399–407. https://www.ncbi.nlm.nih.gov/pubmed/29609831.

Van Dijk, S. J., P. L. Molloy, H. Varinli, J. L. Morrison, B. S. Muhlhausler; Members of EpiSCOPE. "Epigenetics and human obesity." *International Journal of Obesity* 39(1) (January 2015): 85–97. https://www.ncbi.nlm.nih.gov/pubmed/24566855.

Veenendaal, M. V. E., R. C. Painter, S. R. de Rooij, P. M. M. Bossuyt, J. A. M. van der Post, P. D. Gluckman, M. A. Hanson, and T. J. Roseboom. "Transgenerational effects of prenatal exposure to the 1944–45 Dutch famine." *BJOG* 120(5) (April 2013): 548–54. https://obgyn.onlinelibrary.wiley.com/doi/full/10.1111/1471-0528.12136.

Ventimiglia, E., P. Capogrosso, L. Boeri, A. Serino, M. Colicchia, S. Ippolito, R. Scano et al. "Infertility as a proxy of general male health: Results of a cross-sectional survey." *Fertility and Sterility* 104(1) (July 2015): 48–55. https://www.ncbi.nlm.nih.gov/pubmed/26006735.

Wu, H., M. S. Estill, A. Shershebnev, A. Suvorov, S. A. Krawetz, B. W. Whitcomb, H. Dinnie, T. Rahil, C. K. Sites, and J. R. Pilsner. "Preconception urinary phthalate concentrations and sperm DNA methylation profiles among men undergoing IVF treatment: A cross-sectional study." *Human Reproduction* 32(11) (November 2017): 2159–69. https://www.ncbi.nlm.nih.gov/pmc/articles/PMC5850785/.

Yasmin, S. "Experts debunk study that found Holocaust trauma is inherited." *Dallas Morning News*, June 9, 2017. www.chicagotribune.com/lifestyles/health/ct-holocaust-trauma-not-inherited-20170609-story.html.

Yehuda, R., N. P. Daskalakis, A. Lehrner, F. Desarnaud, H. N. Bader, I. Makotkine, J. D. Flory, L. M. Bierer, and M. J. Meaney. "Influences of maternal and paternal PTSD on epigenetic regulation of the glucocorticoid receptor gene in Holocaust survivor offspring." *American Journal of Psychiatry* 171(8) (August 2014): 872–80. https://www.ncbi.nlm.nih.gov/pmc/articles/PMC4127390/.

Chapter Nine:
Imperiling the Planet: *It's Not Just about Humans*

Andrews, G. "Plastics in the ocean affecting human health." Geology and Human Resources. https://serc.carleton.edu/NAGTWorkshops/health/case_studies/plastics.html.

Ankley, G. T., K. K. Coady, M. Gross, H. Holbech, S. L. Levine, G. Maack, and M. Williams. "A critical review of the environmental occurrence and potential effects in aquatic vertebrates of the potent androgen receptor agonist 17ß-trenbolone." *Environmental Toxicology and Chemistry* 37(8) (August 2018): 2064–78. https://www.ncbi.nlm.nih.gov/pmc/articles/PMC6129983/.

Batt, A. L., J. B. Wathen, J. M. Lazorchak, A. R. Olsen, and T. M. Kincaid. "Statistical survey of persistent organic pollutants: Risk estimations to humans and wildlife through consumption of fish from U.S. rivers." *Environmental Science & Technology* 51 (2017): 3021–31. https://digitalcommons.unl.edu/cgi/view content.cgi?article=1262&context=usepapapers.

Bergman, A., J. J. Heindel, S. Jobling, K. A. Kidd, and R. T. Zoeller, eds. *State of the Science of Endocrine Disrupting Chemicals—2012*. World Health Organization, 2013. https://apps.who.int/iris/bitstream/handle/10665/78102/WHO_HSE_PHE_IHE_2013.1_eng.pdf;jsessionid=EFCF73DBEDC17052C00F22B3BD03EBB2?sequence=1.

Davey, J. C., A. P. Nomikos, M. Wungjiranirun, J. R. Sherman, L. Ingram, C. Batki, J. P. Lariviere, and J. W. Hamilton. "Arsenic as an endocrine disruptor: Arsenic disrupts retinoic-acid-receptor- and thyroid-hormone-receptor-mediated gene regulation and thyroid-hormone-mediated amphibian tail metamorphosis." *Environmental Health Perspectives* 116(2) (February 2008): 165–72. https://www.ncbi.nlm.nih.gov/pmc/articles/PMC2235215/.

Edwards, T. M., B. C. Moore, and L. J. Guillette Jr. "Reproductive dysgenesis in wildlife: A comparative view." *International Journal of Andrology* 29(1) (2006): 109–21. https://onlinelibrary.wiley.com/doi/full/10.1111/j.1365-2605.2005.00631.x.

Elliott, J. E., D. A. Kirk, P. A. Martin, L. K. Wilson, G. Kardosi, S. Lee, T. McDaniel, K. D. Hughes, B. D. Smith, and A. M. Idrissi. "Effects of halogenated contaminants on reproductive development in wild mink (*Neovison vison*) from locations in Canada." *Ecotoxicology* 27(5) (July 2018): 539–55. https://www.ncbi.nlm.nih.gov/pubmed/29623614.

Emerson, S. "Human waste is contaminating Australian wildlife with more than 60 pharmaceuticals." Vice.com, November 6, 2018. https://www.vice.com/en_us/article/a3mzve/human-waste-is-contaminating-australian-wildlife-with-more-than-60-pharmaceuticals.

E. O. Wilson Biodiversity Foundation Partners with Art.Science.Gallery. for "Year of the Salamander." Exhibition, March 10, 2014. https://eowilsonfoundation.org/e-o-wilson-biodiversity-foundation-partners-with-art-science-gallery-for-year-of-the-salamander-exhibition/.

EPA. "Persistent organic pollutants: A global issue, a global response." Updated in December 2009. https://www.epa.gov/international-cooperation/persistent-organic-pollutants-global-issue-global-response.

Frederick, P., and N. Jayasena. "Altered pairing behaviour and reproductive success in white ibises exposed to environmentally relevant concentrations of methylmercury." *Proceedings of the Royal Society B: Biological Sciences* 278(1713) (June 22, 2011): 1851–57. https://www.ncbi.nlm.nih.gov/pmc/articles/PMC3097836/.

Georgiou, A. "Mediterranean garbage patch: Huge new 'island' of plastic waste discovered floating in sea." *Newsweek*, May 21, 2019. https://www.newsweek.com/mediterranean-garbage-patch-island-plastic-waste-sea-1431722.

Gibbs, P. E., and G. W. Bryan. "Reproductive failure in populations of the dog-whelk, *Nucella lapillus,* caused by imposex induced by tributyltin from anti-fouling paints." *Journal of the Marine Biological Association of the United Kingdom* 66(4) (November 1986): 767–77. https://www.cambridge.org/core/journals/journal-of-the-marine-biological-association-of-the-united-kingdom/article/reproductive-failure-in-populations-of-the-dogwhelk-nucella-lapillus-caused-by-imposex-induced-by-tributyltin-from-antifouling-paints/091765168341742219A70A9C87FB496E.

Guillette, L. J., Jr., T. S. Gross, G. R. Masson, J. M. Matter, H. Franklin Percival, and A. R. Woodward. "Developmental abnormalities of the gonad and abnormal sex hormone concentrations in juvenile alligators from contaminated and control lakes in Florida." *Environmental Health Perspectives* 102(8) (August 1994): 680–88. https://www.ncbi.nlm.nih.gov/pmc/articles/PMC1567320/.

Hallmann, C. A., M. Sorg, E. Jongejans, H. Siepel, N. Hofland, H. Schwan, W. Stenmans et al. "More than 75 percent decline over 27 years in total flying insect biomass in protected areas." *PLoS One* 12(10) (October 18, 2017): e0185809. https://journals.plos.org/plosone/article?id=10.1371/journal.pone.0185809.

Hui, D. "Food web: Concept and applications." *Nature Education Knowledge* 3(12) (2012): 6. https://www.nature.com/scitable/knowledge/library/food-web-concept-and-applications-84077181/.

Iavicoli, I., L. Fontana, and A. Bergamaschi. "The effects of metals as endocrine disruptors." *Journal of Toxicology and Environmental Health. Part B. Critical Reviews* 12(3) (March 2009): 206–23. https://www.ncbi.nlm.nih.gov/pubmed/19466673.

Jarvis, B. "The insect apocalypse is here." *New York Times Magazine,* November 27, 2018. https://www.nytimes.com/2018/11/27/magazine/insect-apocalypse.html.

Jenssen, B. M. "Effects of anthropogenic endocrine disrupters on responses and adaptations to climate change." In *Endocrine Disrupters,* edited by T. Grotmol, A. Bernhoft, G. S. Eriksen, and T. P. Flaten. Oslo: Norwegian Academy of Science and Letters, 2006. https://pdfs.semanticscholar.org/6211/a40bb3b72ca48c1d0f160575fd5291627e1e.pdf.

Katz, C. "Iceland's seabird colonies are vanishing, with 'massive' chick deaths." *National Geographic,* August 28, 2014. https://www.nationalgeographic.com/news/2014/8/140827-seabird-puffin-tern-iceland-ocean-climate-change-science-winged-warning/.

Kover, P. "Insect 'Armageddon': 5 crucial questions answered." *Scientific American,* October 30, 2017. https://www.scientificamerican.com/article/insect-ldquo-armageddon-rdquo-5-crucial-questions-answered/.

"Let's stop the manipulation of science." *Le Monde,* November 29, 2016. https://www.lemonde.fr/idees/article/2016/11/29/let-s-stop-the-manipulation-of-science_5039867_3232.html.

Lister, B. C., and A. Garcia. "Climate-driven declines in arthropod abundance restructure a rainforest food web." *Proceedings of the National Academy of Sciences* 115(44) (October 30, 2018): E10397–E10406. https://www.pnas.org/con tent/115/44/E10397.

Montanari, S. "Plastic garbage patch bigger than Mexico found in Pacific." *National Geographic*, July 25, 2017. https://www.nationalgeographic.com/news/2017/07 /ocean-plastic-patch-south-pacific-spd/.

Nace, T. "Idyllic Caribbean island covered in a tide of plastic trash along coastline." *Forbes*, October 27, 2017. https://www.forbes.com/sites/trevornace/2017/10/27 /idyllic-caribbean-island-covered-in-a-tide-of-plastic-trash-along-coast line/#6785f46b2524.

Oskam, I. C., E. Ropstad, E. Dahl, E. Lie, A. E. Derocher, O. Wiig, S. Larsen, R. Wiger, and J. U. Skaare. "Organochlorines affect the major androgenic hor mone, testosterone, in male polar bears (*Ursus maritimus*) at Svalbard." *Journal of Toxicology and Environmental Health. Part A.* 66(22) (November 28, 2003): 2119–39. https://www.ncbi.nlm.nih.gov/pubmed/14710596.

Parr, M. "We're losing birds at an alarming rate. We can do something about it." *Washington Post*, September 29, 2019. https://www.washingtonpost.com/opin ions/were-losing-birds-at-an-alarming-rate-we-can-do-something-about -it/2019/09/19/0c25f520-d980-11e9-a688-303693fb4b0b_story.html.

Pelton, E. "Early Thanksgiving counts show a critically low monarch population in California." Xerces Society for Invertebrate Conservation, November 29, 2018. https://xerces.org/2018/11/29/critically-low-monarch-population-in-california/.

Renner, R. "Trash islands are still taking over the oceans at an alarming rate." *Pacific Standard*, March 8, 2018. https://psmag.com/environment/trash-islands -taking-over-oceans.

Rosenberg, M. "Marine life shows disturbing signs of pharmaceutical drug effects." Center for Health Journalism, July 11, 2016. https://www.centerforhealthjour nalism.org/2016/07/16/marine-life-show-disturbing-signs-pharmaceutical -drug-effects.

Schøyen, M., N. W. Green, D. Ø. Hjermann, L. Tveiten, B. Beylich, S. Øxnevad, and J. Beyer. "Levels and trends of tributyltin (TBT) and imposex in dogwhelk (*Nucella lapillus*) along the Norwegian coastline from 1991 to 2017." *Marine Environmental Research* 144 (February 2019): 1–8. https://www.ncbi.nlm.nih .gov/pubmed/30497665.

"Scientists confirm the existence of another ocean garbage patch." ResearchGate, July 19, 2017. https://www.researchgate.net/blog/post/scientists-confirm-the-exist ence-of-another-ocean-garbage-patch.

Stokstad, E. "Zombie endocrine disruptors may threaten aquatic life." *Science* 341(6153) (September 27, 2013): 1441. https://science.sciencemag.org /content/341/6153/1441.

Tomkins, P., M. Saaristo, M. Allinson, and B. B. M. Wong. "Exposure to an agricultural contaminant, 17ß-trenbolone, impairs female mate choice in a freshwater fish." *Aquatic Toxicology* 170 (January 2016): 365–70. https://www.ncbi.nlm.nih.gov/pubmed/26466515.

Chapter Ten:
Imminent Social Insecurities: *Demographic Deviations and the Unraveling of Cultural Institutions*

Batuman, E. "Japan's rent-a-family industry." *New Yorker*, April 23, 2018. https://www.newyorker.com/magazine/2018/04/30/japans-rent-a-family -industry.

"'The best role for women is at home.' Is this the solution to Singapore's falling birth rate?" *Asian Parent*, July 17, 2019. https://sg.theasianparent.com/singapores _falling_birth_rates.

Bricker, D., and J. Ibbitson. *Empty Planet*. New York: Crown, 2019. United Nations: Department of Economics and Social Affairs.

Bruckner, T. A., R. Catalano, and J. Ahern. "Male fetal loss in the U.S. following the terrorist attacks of September 11, 2001." *BMC Public Health* 10(2010): 273. https://www.ncbi.nlm.nih.gov/pmc/articles/PMC2889867/.

Bui, Q., and C. C. Miller. "The age that women have babies: How a gap divides America." *New York Times*, August 4, 2018. https://www.nytimes.com /interactive/2018/08/04/upshot/up-birth-age-gap.html.

del Rio Gomez, I., T. Marshall, P. Tsai, Y. S. Shao, and Y. L. Guo. "Number of boys born to men exposed to polychlorinated biphenyls." *Lancet* 360(9327) (July 13, 2002): 143–44. https://www.ncbi.nlm.nih.gov/pubmed/12126828.

Fukuda, M., K. Fukuda, T. Shimizu, and H. Moller. "Decline in sex ratio at birth after Kobe earthquake." *Human Reproduction* 13(8) (August 1998): 2321–22. https:// www.ncbi.nlm.nih.gov/pubmed/9756319.

Fukuda, M., K. Fukuda, T. Shimizu, M. Nobunaga, L. S. Mamsen, and A. C. Yding. "Climate change is associated with male:female ratios of fetal deaths and newborn infants in Japan." *Fertility and Sterility* 102(5) (November 2014): 1364–70. e2. https://www.ncbi.nlm.nih.gov/pubmed/25226855.

GBD 2017 Population and Fertility Collaborators. "Population and fertility by age and sex for 195 countries and territories, 1950–2017." *Lancet* 392 (November 10, 2018): 1995–2051. https://www.thelancet.com/journals/lancet/article/ PIIS0140-6736(18)32278-5/fulltext.

Hamilton, B. E., J. A. Martin, M. J. K. Osterman, and L. M. Rossen. "Births: Provisional data for 2018." Report No. 007, May 2019. US Department of Health and Human Services, Centers for Disease Control and Prevention, National Center for Health Statistics, National Vital Statistics System. https://www.cdc .gov/nchs/data/vsrr/vsrr-007-508.pdf.

Hay, M. "Why are the Japanese still not fucking?" Vice.com, January 22, 2015. https://www.vice.com/da/article/7b7y8x/why-arent-the-japanese-fucking-361.

"Japan's problem with celibacy and sexlessness." Breaking Asia, March 19, 2019. https://www.breakingasia.com/360/japans-problem-with-celibacy-and -sexlessness/.

Jozuka, E. "Inside the Japanese town that pays cash for kids." CNN Health, February 3, 2019. https://www.cnn.com/2018/12/27/health/japan-fertility-birth-rate -children-intl/index.html.

Lutz, W., V. Skirbekk, and M. R. Testa. "The low-fertility trap hypothesis: Forces that may lead to further postponement and fewer births in Europe." International Institute for Applied Systems Analysis, RP-07-001, March 2007. http://pure .iiasa.ac.at/id/eprint/8465/1/RP-07-001.pdf.

Mather, M., L. A. Jacobsen, and K. M. Pollard. "Aging in the United States." *Population Bulletin* 70 (2) (December 2015). Population Reference Bureau. https://www.prb .org/wp-content/uploads/2016/01/aging-us-population-bulletin-1.pdf.

Meola, A. "The aging population in the US is causing problems for our health-care costs." *Business Insider*, July 18, 2019. https://www.businessinsider.com /aging-population-healthcare.

Moore, C. "The village of the dolls: Artist creates mannequins and leaves them around her village in Japan as the local population dwindles." DailyMail .com, April 22, 2016. https://www.dailymail.co.uk/news/article-3553992 /The-village-dolls-Artist-creates-mannequins-leaves-village-Japan-local -population-dwindles.html.

National Institute of Population and Social Security Research. "Population Projections for Japan (2016–2065)." http://www.ipss.go.jp/pp-zenkoku/e/zenkoku _e2017/pp_zenkoku2017e_gaiyou.html.

Obel, C., T. B. Henriksen, N. J. Secher, B. Eskenazi, and M. Hedegaard. "Psychological distress during early gestation and offspring sex ratio." *Human Reproduction* 22(11) (November 2007): 3009–12. https://www.ncbi.nlm.nih.gov /pubmed/17768170.

Parker, K., J. M. Horowitz, A. Brown, R. Fry, D. V. Cohn, and R. Igielnik. "Demographic and economic trends in urban, suburban and rural communities." Pew Research Center, May 22, 2018. https://www.pewsocialtrends.org/2018/05/22/demographic -and-economic-trends-in-urban-suburban-and-rural-communities/.

Pavic, D. "A review of environmental and occupational toxins in relation to sex ratio at birth." *Early Human Development* 141 (February 2020): 104873. https://www.ncbi .nlm.nih.gov/pubmed/31506206.

Perlberg, S. "World population will peak in 2055 unless we discover the 'elixir of immortality.'" *Business Insider*, September 9, 2013. https://www.businessinsider .com/deutsche-population-will-peak-in-2055-2013-9.

Pettit, C. "Countries where people have the most and least sex." Weekly Gravy,

May 20, 2014. https://weeklygravy.com/lifestyle/countries-where-people-have -the-most-and-least-sex/.

Pew Research Center. "Population change in the US and the world from 1950 to 2015." January 30, 2014. https://www.pewresearch.org/global/2014/01/30/chapter -4-population-change-in-the-u-s-and-the-world-from-1950-to-2050/.

Pradhan, E. "Female education and childbearing: A closer look at the data." *World Bank Blogs*, November 24, 2015. https://blogs.worldbank.org/health/female-ed ucation-and-childbearing-closer-look-data.

Prosser, M. "Searching for a cure for Japan's loneliness epidemic." HuffPost, August 15, 2018. https://www.huffpost.com/entry/japan-loneliness-aging -robots-technology_n_5b72873ae4b0530743cd04aa.

Randers, *Earth in 2052*. TEDxTrondheimSalon, 2014. https://www.youtube .com/watch?v=gPEVfXVyNMM.

Sin, Y. "Govt aid alone not enough to raise birth rate: Minister." *Straits Times*, March 2, 2018. https://www.straitstimes.com/singapore/govt-aid-alone-not-enough -to-raise-birth-rate-minister.

"6 reasons why the Japanese aren't having babies." YouTube. https://www.youtube .com/watch?v=4pXSJ35_v2M.

Stritof, S. "Estimated median age of first marriage by gender: 1890 to 2018." The Spruce, December 1, 2019. https://www.thespruce.com/estimated-median-age -marriage-2303878.

"The 2017 annual report of the Board of Trustees of the Federal Old-Age and Sur- vivors Insurance and Federal Disability Insurance Trust Funds." July 13, 2017. https://www.ssa.gov/oact/TR/2017/tr2017.pdf.

United States Census Bureau. "Older people projected to outnumber children for first time in U.S. history." October 8, 2019. https://www.census.gov/newsroom /press-releases/2018/cb18-41-population-projections.html.

University of Melbourne. "Women's choice drives more sustainable global birth rate." Futurity, November 1, 2018. https://www.futurity.org/global-fertil ity-rates-1901352/.

Waldman, K. "The XX factor: Young people in Japan have given up on sex." Slate, October 22, 2013. https://slate.com/human-interest/2013/10/celibacy-syn drome-in-japan-why-aren-t-young-people-interested-in-sex-or-relationship s.html.

Wee, S-L., and S. L. Myers. "China's birthrate hits historic low, in looming crisis for Beijing." *New York Times*, January 16, 2020. https://www.nytimes .com/2020/01/16/business/china-birth-rate-2019.html.

World Bank. "Fertility rate, total (births per woman)." 2019. https://data.worldbank .org/indicator/SP.DYN.TFRT.IN.

"World Population Prospects 2019." https://population.un.org/wpp/Graphs/Prob abilistic/POP/TOT/900.

Chapter Eleven:
A Personal Protection Plan: *Cleaning Up Our Harmful Habits*

Al-Jaroudi, D., N. Al-Banyan, N. J. Aljohani, O. Kaddour, and M. Al-Tannir. "Vitamin D deficiency among subfertile women: Case-control study." *Gynecological Endocrinology* 32(4) (December 11, 2016): 272–75. https://www.ncbi.nlm.nih.gov/pubmed/?term=26573125.

Bae, J., S. Park, and J-W. Kwon. "Factors associated with menstrual cycle irregularity and menopause." *BMC Women's Health* 18 (February 6, 2018): 36. https://www.ncbi.nlm.nih.gov/pmc/articles/PMC5801702/.

Best, D., A. Avenell, and S. Bhattacharya. "How effective are weight-loss interventions for improving fertility in women and men who are overweight or obese? A systematic review and meta-analysis of the evidence." *Human Reproduction Update* 23(6) (November 1, 2017): 681–705. https://www.ncbi.nlm.nih.gov/pubmed/28961722.

Cito, G., A. Cocci, E. Micelli, A. Gabutti, G. I. Russo, M. E. Coccia, G. Franco, S. Serni, M. Carini, and A. Natali. "Vitamin D and male fertility: An updated review." *World Journal of Men's Health* 38(2) (May 17, 2019): 164–77. https://wjmh.org/DOIx.php?id=10.5534/wjmh.190057.

Efrat, M., A. Stein, H. Pinkas, R. Unger, and R. Birk. "Dietary patterns are positively associated with semen quality." *Fertility and Sterility* 109(5) (May 2018): 809–16. https://www.fertstert.org/article/S0015-0282(18)30010-4/fulltext.

"EWG's Consumer guide to seafood." https://www.ewg.org/research/ewgs-good-seafood-guide/executive-summary.

Gaskins, A. J., J. Mendiola, M. Afeiche, N. Jørgensen, S. H. Swan, and J. E. Chavarro. "Physical activity and television watching in relation to semen quality in young men." *British Journal of Sports Medicine* 49(4) (February 4, 2013): 265–70. https://www.ncbi.nlm.nih.gov/pmc/articles/PMC3868632/.

Gudmundsdottir, S. L., W. D. Flanders, and L. B. Augestad. "Physical activity and fertility in women: The North-Trøndelag Health Study." *Human Reproduction* 24(12) (October 3, 2009): 3196–204. https://academic.oup.com/humrep/article/24/12/3196/647657.

Jalali-Chimeh, F., A. Gholamrezaei, M. Vafa, M. Nasiri, B. Abiri, T. Darooneh, and G. Ozgoli. "Effect of vitamin D therapy on sexual function in women with sexual dysfunction and vitamin D deficiency: A randomized, double-blind, placebo controlled clinical trial." *Journal of Urology* 201(5) (May 2019): 987–93. https://www.auajournals.org/doi/10.1016/j.juro.2018.10.019.

Jensen, T. K., L. Priskorn, S. A. Holmboe, F. L. Nassan, A-M. Andersson, C. Dalgård, J. Holm Petersen, J. E. Chavarro, and N. Jørgensen. "Associations of fish oil supplement use with testicular function in young men." *JAMA Network Open* 3(1) (January 17, 2020). https://jamanetwork.com/journals/jamanetworkopen

/fullarticle/2758861?widget=personalizedcontent&previousaticle=2758855#editori
al-comment-tab.

Karayiannis, D., M. D. Kontogianni, C. Mendorou, L. Douka, M. Mastrominas, and
N. Yiannakouris. "Association between adherence to the Mediterranean diet
and semen quality parameters in male partners of couples attempting fertility."
Human Reproduction 32(1) (January 1, 2017): 215–22. https://academic.oup
.com/humrep/article/32/1/215/2513723.

Li, J., L. Long, Y. Liu, W. He, and M. Li. "Effects of a mindfulness-based intervention on fertility quality of life and pregnancy rates among women subjected to first in vitro fertilization treatment." *Behaviour Research Therapy* 77
(February 2016): 96–104. https://www.sciencedirect.com/science/article/abs/pii
/S0005796715300747.

Luque, E. M., A. Tissera, M. P. Gaggino, R. I. Molina, A. Mangeaud, L. M. Vincenti,
F. Beltramone et al. "Body mass index and human sperm quality: Neither one
extreme nor the other." *Reproduction, Fertility and Development* 29(4) (December 18, 2015): 731–39. https://www.publish.csiro.au/rd/RD15351.

Natt, D., U. Kugelberg, E. Casas, E. Nedstrand, S. Zalavary, P. Henriksson, C.
Nijm et al. "Human sperm displays rapid responses to diet." *PLoS Biology*
17(12) (December 26, 2019). https://www.ncbi.nlm.nih.gov/pmc/articles
/PMC6932762/pdf/pbio.3000559.pdf.

Orio, F., G. Muscogiuri, A. Ascione, F. Marciano, A. Volpe, G. La Sala, S. Savas-
tano, A. Colao, S. Palomba, and S. Minerva. "Effects of physical exercise on the
female reproductive system." *Endocrinology* 38(3) (September 2013): 305–19.
https://www.ncbi.nlm.nih.gov/pubmed/24126551.

Park, J., J. B. Stanford, C. A. Porucznik, K. Christensen, and K. C. Schliep.
"Daily perceived stress and time to pregnancy: A prospective cohort
study of women trying to conceive." *Psychoneuroendocrinology* 110 (December 2019): 104446. https://www.sciencedirect.com/science/article/abs/pii
/S0306453019303932.

Ramaraju, G. A., S. Teppala, K. Prathigudupu, M. Kalagara, S. Thota, and R. Cheemakurthi. "Association between obesity and sperm quality." *Andrologia* 50(3)
(September 19, 2017). https://onlinelibrary.wiley.com/doi/abs/10.1111/
and.12888.

Rampton, J. "20 Quotes from Jim Rohn putting success and life into perspective."
Entrepreneur, March 4, 2016. https://www.entrepreneur.com/article/271873.

Ricci, E., S. Noli, S. Ferrari, I. La Vecchia, M. Castiglioni, S. Cipriani, F. Parazzini,
and C. Agostoni. "Fatty acids, food groups and semen variables in men referring to an Italian fertility clinic: Cross-sectional analysis of a prospective cohort
study." *Andrologia* 52(3) (January 8, 2020): e13505. https://www.ncbi.nlm.nih
.gov/pubmed/31912922.

Rosety, M. Á., A. J. Díaz, J. M. Rosety, M. T. Pery, F. Brenes-Martín, M. Bernardi, N.

García, M. Rosety-Rodríguez, F. J. Ordoñez, and I. Rosety. "Exercise improved semen quality and reproductive hormone levels in sedentary obese adults." *Nutricion Hospitalaria* 34(3) (June 5, 2017): 603–7. https://www.ncbi.nlm.nih.gov/pubmed/28627195.

Russo, L. M., B. W. Whitcomb, L. Sunni, L. Mumford, M. Hawkins, R. G. Radin, K. C. Schliep et al. "A prospective study of physical activity and fecundability in women with a history of pregnancy loss." *Human Reproduction.* 33(7) (April 10, 2018): 1291–98. https://www.ncbi.nlm.nih.gov/pmc/articles/PMC6012250/pdf/dey086.pdf.

Salas-Huetos, A., M. Bulló, and J. Salas-Salvadó. "Dietary patterns, foods and nutrients in male fertility parameters and fecundability: A systematic review of observational studies." *Human Reproduction Update* 23(4) (July 1, 2017): 371–89. https://www.ncbi.nlm.nih.gov/pubmed/28333357.

Silvestris, E., D. Lovero, and R. Palmirotta. "Nutrition and female fertility: An interdependent correlation." *Frontiers in Endocrinology* (Lausanne, Switzerland) 10 (June 7, 2019): 346. https://www.ncbi.nlm.nih.gov/pmc/articles/PMC6568019/.

"Smoking and infertility." Fact sheet. *American Society for Reproductive Medicine*, 2014. https://www.reproductivefacts.org/globalassets/rf/news-and-public ations/bookletsfact-sheets/english-fact-sheets-and-info-booklets/smoking _and_infertility_factsheet.pdf.

Sun, B., C. Messerlian, Z. H. Sun, P. Duan, H. G. Chen, Y. J. Chen, P. Wang et al. "Physical activity and sedentary time in relation to semen quality in healthy men screened as potential sperm donors." *Human Reproduction* 34(12) (December 1, 2019): 2330–39. https://www.ncbi.nlm.nih.gov/pubmed/31858122.

Toledo, E., C. López-del Burgo, A. Ruiz-Zambrana, M. Donazar, Í. Navarro-Blasco, M. A. Martínez-González, and J. de Irala. "Dietary patterns and difficulty conceiving: A nested case-control study." *Fertility and Sterility* 96(2011): 1149–53. https://www.sciencedirect.com/science/article/abs/pii/S001502821102485X.

Vujkovic, M., J. H. de Vries, J. Lindemans, N. S. Macklon, P. J. van der Spek, E. A. Steegers, and R. P. Steegers-Theunissen. "The preconception Mediterranean dietary pattern in couples undergoing in vitro fertilization/intracytoplasmic sperm injection treatment increases the chance of pregnancy." *Fertility and Sterility* 94(6) (November 2010): 2096–101. https://www.ncbi.nlm.nih.gov/pubmed/?term=20189169.

Wells, D. "Sauna and pregnancy: Safety and risks." *Healthline: Parenthood*, July 21, 2016. https://www.healthline.com/health/pregnancy/sauna.

Chapter Twelve:
Reducing the Chemical Footprint in Your Home: *Making It a Safer Haven*

American Chemical Society. "Keep off the grass and take off your shoes! Common sense can stop pesticides from being tracked into the house." ScienceDaily, April

27, 1999. https://www.sciencedaily.com/releases/1999/04/990427045111 .htm.

"Cosmetics, body care products, and personal care products." National Organic Program, April 2008. https://www.ams.usda.gov/sites/default/files/media/Organ icCosmeticsFactSheet.pdf.

Food and Water Watch. "Understanding food labels." July 12, 2018. https://www .foodandwaterwatch.org/about/live-healthy/consumer-labels.

Hagen, L. "Natural method to get rid of common garden weeds." *Garden Design.* https://www.gardendesign.com/how-to/weeds.html.

Harley, K. G., K. Kogut, D. S. Madrigal, M. Cardenas, I. A. Vera, G. Meza-Alfaro, J. She, Q. Gavin, R. Zahedi, A. Bradman, B. Eskenazi, and K. L. Parra. "Reducing phthalate, paraben, and phenol exposure from personal care products in adolescent girls: Findings from the HERMOSA Intervention Study." *Environmental Health Perspectives* 124(10) (October 2016): 1600–1607. https://www.ncbi.nlm .nih.gov/pmc/articles/PMC5047791/.

Healthy Stuff. "New study rates best and worst garden hoses: Lead, phthalates & hazardous flame retardants in garden hoses." Ecology Center, June 20, 2016. https://www.ecocenter.org/healthy-stuff/new-study-rates-best-and-worst-gar den-hoses-lead-phthalates-hazardous-flame-retardants-garden-hoses.

Hyland, C., A. Bradman, R. Gerona, S. Patton, I. Zakharevich, R. B. Gunier, and K. Klein. "Organic diet intervention significantly reduces urinary pesticide levels in U.S. children and adults." *Environmental Research* 171 (April 2019): 568–75. https://www.sciencedirect.com/science/article/pii/S0013935119300246.

"Inert ingredients of pesticide products." Environmental Protection Agency, October 10, 1989. https://www.epa.gov/sites/production/files/2015-10/documents /fr54.pdf.

Kinch, C. D., K. Ibhazehiebo, J. H. Jeong, H. R. Habib, and D. M. Kurrasch. "Low-dose exposure to bisphenol A and replacement bisphenol S induces precocious hypothalamic neurogenesis in embryonic zebrafish." *Proceedings of the National Academy of Sciences USA* 112(5) (February 3, 2015): 1475–80. https://www.ncbi .nlm.nih.gov/pubmed/25583509.

Koch, H. M., M. Lorber, K. L. Christensen, C. Pälmke, S. Koslitz, and T. Brüning. "Identifying sources of phthalate exposure with human biomonitoring: Results of a 48h fasting study with urine collection and personal activity patterns." *International Journal of Hygiene and Environmental Health* 216(6) (November 2013): 672–81. https://www.sciencedirect.com/science/article/abs/pii /S1438463912001381.

Mitro, S. D., R. E. Dodson, V. Singla, G. Adamkiewicz, A. F. Elmi, M. K. Tilly, A. R. Zota. "Consumer product chemicals in indoor dust: A quantitative meta-analysis of U.S. studies." *Environmental Science & Technology* 50(19) (October 4, 2016): 10661–72. https://www.ncbi.nlm.nih.gov/pmc/articles/PMC5052660/.

"Naphthalene: Technical fact sheet." National Pesticide Information Center, Oregon State University. http://npic.orst.edu/factsheets/archive/naphtech.html.

Stoiber, T. "What are parabens, and why don't they belong in cosmetics?" Environmental Working Group, April 9, 2019. https://www.ewg.org/californiacosmetics/parabens.

US Food and Drug Administration. "Where and how to dispose of unused medicines." March 11, 2020. https://www.fda.gov/consumers/consumer-updates/where-and-how-dispose-unused-medicines.

Varshavsky, J. R., R. Morello-Frosch, T. J. Woodruff, and A. R. Zota. "Dietary sources of cumulative phthalate exposure among the U.S. general population in NHANES 2005–2014." *Environment International* 115 (June 2018): 417–29. https://www.ncbi.nlm.nih.gov/pmc/articles/PMC5970069/.

Chapter Thirteen:
Envisioning a Healthier Future: *What Needs to Be Done*

Allen, J. "Stop playing whack-a-mole with hazardous chemicals." *Washington Post*, December 15, 2016. https://www.washingtonpost.com/opinions/stop-playing-whack-a-mole-with-hazardous-chemicals/2016/12/15/9a357090-bb36-11e6-91ee-1adddfe36cbe_story.html.

Bornehag, C. G., F. Carlstedt, B. A. G. Jönsson, C. H. Lindh, T. K. Jensen, A. Bodin, C. Jonsson, S. Janson, and S. H. Swan. "Prenatal phthalate exposures and anogenital distance in Swedish boys." *Environmental Health Perspectives* 123(1) (January 2015): 101–7. https://www.ncbi.nlm.nih.gov/pmc/articles/PMC4286276/.

Constable, P. "Pakistan moves to ban single-use plastic bags: 'The health of 200 million people is at stake.'" *Washington Post*, August 13, 2019. https://www.washingtonpost.com/world/asia_pacific/pakistan-moves-to-ban-single-use-plastic-bags-the-health-of-200-million-people-is-at-stake/2019/08/12/6c7641ca-bc23-11e9-b873-63ace636af08_story.html.

"CPSC prohibits certain phthalates in children's toys and child care products." US Consumer Product Safety Commission, October 20, 2017. https://www.cpsc.gov/Newsroom/News-Releases/2018/CPSC-Prohibits-Certain-Phthalates-in-Childrens-Toys-and-Child-Care-Products.

Editor. "Endangered animals saved from extinction." All About Wildlife, May 16, 2011.

"Enhancing sustainability." Walmart.org, 2020. https://walmart.org/what-we-do/enhancing-sustainability.

"Green chemistry." Wikipedia. https://en.wikipedia.org/wiki/Green_chemistry.

"The Home Depot announces to stop selling carpets treated with toxic stain-resistant PFAS chemicals." Environmental Defence and Safer Chemicals, Healthy Families, September 18, 2019. https://environmentaldefence.ca/2019/09/18/home-depot-announces-stop-selling-carpets-treated-toxic-stain-resistant-pfas-chemicals/.

"Is oxybenzone contributing to the death of coral reefs?" SunscreenSafety.info. https://www.sunscreensafety.info/oxybenzone-coral-reefs/.

Li, D-K., Z. Zhou, M. Miao, Y. He, J-T. Wang, J. Ferber, L. J. Herrinton, E-S. Gao, and W. Yuan. "Urine bisphenol-A (BPA) level in relation to semen quality." *Fertility and Sterility* 95(2) (February 2011): 625–30.e1-4. https://pubmed.ncbi.nlm.nih.gov/21035116/.

Perara, F., J. Vishnevetsky, J. B. Herbstman, A. M. Calafat, W. Xiong, V. Rauh, and S. Wang, "Prenatal bisphenol A exposure and child behavior in an inner-city cohort." *Environmental Health Perspectives* 120 (8) (August 2012): 1190–94. https://pubmed.ncbi.nlm.nih.gov/22543054/.

REACH. European Commission, July 8, 2019. https://ec.europa.eu/environment/chemicals/reach/reach_en.htm.

"Species directory." World Wildlife Fund. https://www.worldwildlife.org/species/directory?sort=extinction_status&direction=desc.

Stanton, R. L., C. A. Morrissey, and R. G. Clark. "Analysis of trends and agricultural drivers of farmland bird declines in North America: A review." *Agriculture, Ecosystems & Environment* 254 (February 15, 2018): 244–54. https://www.sciencedirect.com/science/article/abs/pii/S016788091730525X.

Steffen, A. D. "Australia came up with a way to save the oceans from plastic pollution and garbage." *Intelligent Living,* February 10, 2019. https://www.intelligentliving.co/australia-plastic-ocean/.

Steffen, L. "Costa Rica set to become the world's first plastic-free and carbon-free country by 2021." *Intelligent Living,* May 10, 2019. https://www.intelligentliving.co/costa-rica-plastic-carbon-free-2021/.

Vandenbergh, L. N. "Non-monotonic dose responses in studies of endocrine disrupting chemicals: Bisphenol A as a case study." *Dose Response* 12(2) (May 2014): 259–76. https://www.ncbi.nlm.nih.gov/pmc/articles/PMC4036398/.

Vandenberg, L. N., T. Colborn, T. B. Hayes, J. J. Heindel, D. R. Jacobs Jr., D-H. Lee, T. Shioda et al. "Hormones and endocrine-disrupting chemicals: Low-dose effects and nonmonotonic dose responses." *Endocrine Reviews* 33(3) (June 2012): 378–455. https://www.ncbi.nlm.nih.gov/pmc/articles/PMC3365860/.

"Walmart releases high priority chemical list." ChemicalWatch, 2020. https://chemicalwatch.com/48724/walmart-releases-high-priority-chemical-list.

Watson, A. "Companies putting public health at risk by replacing one harmful chemical with similar, potentially toxic, alternatives." CHEM Trust, March 27, 2018. https://chemtrust.org/toxicsoup/.

"Wingspread Conference on the Precautionary Principle." Science & Environmental Health Network, January 26, 1998. https://www.sehn.org/sehn/wingspread-conference-on-the-precautionary-principle.

Conclusion

Anonymous. "Thirty years of a smallpox-free world." College of Physicians of Philadelphia, May 8, 2010. https://www.historyofvaccines.org/content/blog/thirty-years-smallpox-free-world.

"Clean Air Act overview." US Environmental Protection Agency. https://www.epa.gov/clean-air-act-overview/progress-cleaning-air-and-improving-peoples-health.

"Diseases you almost forgot about (thanks to vaccines)." Centers for Disease Control and Prevention, January 3, 2020. https://www.cdc.gov/vaccines/parents/diseases/forgot-14-diseases.html.

Pirkle, J. L., D. J. Brody, E. W. Gunther et al. "The decline in blood lead levels in the United States." *JAMA*, July 27, 1994. https://jamanetwork.com/journals/jama/article-abstract/376894.

"Report: Cleaning up Great Lakes boosts economic development," *Grand Rapids Business Journal*, August 13, 2019.

Whorton, M. D., and T. H. Milby. "Recovery of testicular function among DBCP workers." *Journal of Occupational Medicine* 22 (3) (March 1980): 177–79. https://pubmed.ncbi.nlm.nih.gov/7365555/.

INDEX

ABOUT DR. SHANNA SWAN

Shanna H. Swan, PhD, is one of the world's leading environmental and reproductive epidemiologists and a professor of environmental medicine and public health at the Icahn School of Medicine at Mount Sinai in New York City. An award-winning scientist, her work examines the impact of environmental exposures, including chemicals such as phthalates and bisphenol A, on men's and women's reproductive health and the neurodevelopment of children.

For more than twenty years, Dr. Swan and her colleagues have been studying the dramatic decline in sperm count around the world and the impact of environmental chemicals and pharmaceuticals on reproductive tract development and neurodevelopment. Her July 2017 paper "Temporal Trends in Sperm Count: A Systematic Review and Meta-Regression Analysis" ranked #26 among all referenced scientific papers published in 2017 worldwide.

Dr. Swan has published more than two hundred scientific papers and myriad book chapters and has been featured in extensive media coverage around the world. Her appearances include *ABC News*, *NBC Nightly News*, *60 Minutes*, *CBS News*, PBS, the BBC, PRI, and NPR, as well as in leading magazines and newspapers, ranging from the *Washington Post* to *Bloomberg News* to *New Scientist*. You can find her at shannaswan.com.

ABOUT STACEY COLINO

Stacey Colino is an award-winning writer, specializing in science, health, and psychological issues. She is a regular contributor to *U.S. News & World Report* and AARP.org, and her work has also appeared in the *Washington*

Post Health and Wellness sections and in dozens of national magazines, including *Prevention, Health, Newsweek, Parade, Cosmopolitan, Real Simple, Woman's Day, Good Housekeeping, Harper's Bazaar, Marie Claire, Shape, Parents,* and *Men's Journal.* She has collaborated on more than a dozen books and is the coauthor of *Emotional Inflammation: Discover Your Triggers and Reclaim Your Equilibrium During Anxious Times* with Lise Van Susteren, MD (Sounds True, 2020); *Disease-Proof: The Remarkable Truth About What Makes Us Well* with David L. Katz, MD (Hudson Street Press, 2013); and *Taking Back the Month: A Personalized Solution for Managing PMS and Enhancing Your Health* with Diana Taylor, PhD (Perigee, 2002).